GRANDPA
GOT HOOKED!

Fill in Books with **70** Drills

Crossword Puzzles for Dementia Patients

PUZZLE THERAPIST
CROSSWORD | SUDOKU | KIDS & ADULTS

CONTENT

PUZZLE 1

ACROSS

1. African capital

6. '___ Breaky Heart"

10. Prefix for scope

14. Did a smithy's job

15. Mass in Arctic waters

16. Jack-in-the-pulpit, e.g.

17. Pleasant greeting

19. Voyeur's look-see

20. Bitty bark

21. Chowderhead

22. Flowering month

24. Media magnate Murdoch

26. Onslaught

30. Cyrano's prominent feature

31. Unstructured consciousness

32. Word with animal or a punch

34. "And let us not be ___ in well doing" (Gal. 6:9)

35. Ole's kin

36. Word with belly or ear

37. Fine partner

38. Glutton's request

39. Stubbed item

40. Spacek of the screen

41. Pre-1991 superpower

42. Nettle

44. Divine circle

46. Before the footlights

47. Stereotypes

49. Classification system for blood

50. Window base

51. Prickly seed vessel (Var.)

53. It'll hold water

56. Irritation or annoyance

59. Fruit with a wrinkled rind

60. Asian housemaid

61. River past Amiens

62. Powder substance

63. Winter Olympics event

64. Son of Japheth

DOWN

1. Covered with soot, e.g.

2. Mormon Tabernacle, for one

3. Horseback

4. 'Curse you, ___ Baron!"

5. Sun-dried bricks

6. Currently in progress

7. Layered sandwich

8. Masonry trough

9. Freeholders

10. Papyrus plant

11. Poet's "before"

12. Feel sorry about

13. Sitter's handful

18. Knight time?

23. Aesthetically pretentious

25. Jab playfully

26. Like some eyes

27. LBO for an exterminator?

28. Growls

29. Anesthetic of yore

31. Common thing

32. Place for a grilling

33. Chipmunk snack

34. Use inefficiently, as time

37. Slash

38. Backless slipper

40. Cutting thrust

43. Slanted, as type

44. Reception site, perhaps

45. Superlatively capable

47. Gracefully limber

48. Cum laude start

50. Dateless

52. Film spool

53. Word with up or down

54. Turkish title

55. Well, just the opposite?

57. Aussie avian

58. Old Cannes coin

PUZZLE 2

ACROSS

1. One to hang with

4. Works underwater

11. Grain husk

16. Santa ___, Calif.

17. Stud locale

18. Fox hunter's cry

19. Serve as a commentator, as in sportscasting

21. Hole-making tool

22. Forbidden perfume?

23. Dead expanse

24. Typical of one's sex

25. Religious song

31. Overloaded at the top

34. 'That's all ___ wrote"

35. Blood pressure raiser

36. Classification system for blood

37. What gives a pose poise?

38. Type of letter

41. A female actor

44. Ham sandwich bread, often

45. Maintained

46. Freshwater fishes

48. Gherkin kin

52. Baby sound

55. Television, radio, print, etc.

58. Impulse to travel

62. 'Dead man's hand" card

63. The three to an inning?

64. Howe'er

65. Letters on a motor-oil can

66. Ivory ticklers

68. Two cotyledons in the seed

72. Drinks to excess

73. Poetic nightfall

74. One of the deadly sins

78. Mormon Tabernacle, for one

79. An actress

83. Pasta possibility

84. Wing part

85. Like some ears

86. Bothered to do something

87. Runs out on

88. Be decisive

DOWN

1. Formal agreement

2. Endangered buffalo

3. Nursery rhyme pet

4. Cover charge, for example

5. Bled

6. Table scrap

7. Smooth

8. Composition with sacred lyrics

9. Nautical position

10. Homer's neighbor

11. Jurisdiction in equity

12. African ethnic group

13. Birthplace of Camus

14. A skeletal muscle

15. The "F" in FYI

20. 'How ___ is that doggie in the window?"

24. Theater guide?

26. Beer topper

27. Rampur royalty

28. Like Harvard's walls

29. Egyptian cobra

30. 'Enter the Dragon" star

31. Corkboard item

32. Double-reed woodwind

33. Majestic display

38. Not edited

39. Emulate old paint

40. Relinquish control

42. Inscribe for good

43. Nominate

47. Laundry problem

49. March 15, historically

50. Pocket collection

51. Highland youth

53. Worked the Delaware Bay, perhaps

54. A white person

56. Civil aviation

57. Carte before the course

58. '___ be my pleasure!"

59. Away from the speaker

60. Insects to protect pupas

61. Japanese delicacy

66. Soft thin cloth woven

67. Anagram of "lies"

69. State one's views

70. Like poltergeists

71. Does a casino job

75. A place ___ itself

76. Editing room sound

77. Circus structure

78. Bookkeeper

79. Small amount

80. Leave the straight and narrow

81. One of three in Fiji

82. Those elected

PUZZLE 3

1	2	3	4		5	6	7	8	9	10		11	12	13	14	15
16					17							18				
19			20									21				
22					23					24						
25				26				27	28							
29				30		31	32									
	33	34	35						36							
37	38				39		40				41	42	43	44		
45				46				47	48							
49				50				51								
		52	53			54		55								
56	57	58	59			60					61	62	63			
64							65		66	67						
68				69	70	71		72								
73				74			75									
76				77					78							
79				80				81								

ACROSS

1. "Buddenbrooks" author

5. Valley

11. Orbital point

16. Dwarf buffalo of Indonesia

17. Astronomy Muse

18. Life force in yoga

19. Low-spirited

21. Pertaining to blood

22. Abreast (of)

23. Blubber

24. Focal point

25. 20 Questions category

27. Cut off

29. Blue

30. Revolving drum

33. Alexander's rule, e.g.

36. Assent

37. Cleave

39. "How ___ Mehta Got Kissed, Got Wild, and Got a Life" (Kaavya Viswanathan novel in the news)

41. Canaanite deity

45. The most important dish

49. Camping gear

50. "Frasier" actress Gilpin

51. Freshen

52. "A pox on you!"

54. Idiosyncrasy

56. Depart hastily

61. Cooking meas.

64. Caution

65. Grace's prince

68. Canaries' cousins

69. Need a bath badly

72. The "N" of U.N.C.F.

73. Photographer's request

74. Between first and last name

76. Fast finisher?

77. Spinners in a blender

78. Doctor Who villainess, with "the"

79. Autocrats

80. Nearly

81. "Let it stand"

DOWN

1. House keepers

2. Sightlessness

3. Once in a while

4. "Eraserhead" star Jack

5. "So ___ me!"

6. Fished with a net

7. Pink, as a steak

8. A chip, maybe

9. Haberdashery item

10. "We've been ___!"

11. Point in the orbit of a planet or comet

12. Kind of income

13. Silk fabric

14. Comparatively cockamamie

15. Courses with greens

20. "___ Town Too" (1981 hit)

24. "Coriolanus" setting

26. Owning land

28. Needles

31. Bond player

32. Kind of penguin

34. The Amish, e.g.

35. "___ we having fun yet?"

37. Appropriate

38. Conk out

40. "All kidding ___..."

41. Blocks

42. Compound lens or lens

system

43. Appear

44. "Malcolm X" director

46. Fencing equipment

47. Cupid's boss

48. Athletic supporter?

52. Fans

53. Western blue flag, e.g.

55. To grow dark

56. Droopy eared hound

57. Plant swellings

58. Hindu drink of the gods

59. Tillie of 1920s comics

60. Tickle pink

62. Calm

63. "To your health!"

66. Setting for TV's "Newhart"

67. Approaches

70. ___ cheese

71. Taro root

74. "Dilbert" cartoonist Scott Adams has one: Abbr.

75. "C'___ la vie!"

PUZZLE 4

1	2	3	4			5	6	7	8	9	10	11		12	13	14
15				16		17								18		
19						20						21				
22				23							24					
25				26				27	28	29						
	30		31				32	33				34				
		35		36						37						
38	39	40			41				42		43	44	45	46		
47				48	49			50	51							
52						53				54						
		55		56	57			58	59							
60	61	62	63		64				65		66	67				
68				69			70				71	72				
73					74	75			76							
77				78	79	80			81							
82			83					84								
85			86					87								

ACROSS

1. Marinate

5. Covenant recipient (Gen. 17:9)

12. Tach reading

15. Or located near a hilum

17. In an unstructured way

18. Romantic or Victorian, e.g.

19. Homer's first letter

20. Wet lightly

22. Evaluate the grounds

for indictments

24. Absinthe flavoring

25. 'Not my error" notation

26. ___ Perignon

27. Review in detail

30. Scottish Celts, e.g.

32. Slightly inclined

34. Dampens

35. Oval seeds

38. They're history

41. It's Big in London

42. Yeltsin's bailiwick

47. Two or more computers or dedicated

52. Used for transactions

53. Ovoid objects, to Romans

54. Approximately

55. Machine or vehicle

60. Custard base

64. Necessary things

65. Echo, e.g.

68. Type of politics

70. Lush surroundings?

71. Bravo or Lobo

73. ___ artery

74. Trailblazed

77. holds the copy for the compositor

81. Message boat

82. 'Human Concretion" artist

83. Relating to or suggestive

84. Ohio or Missouri

85. 'Enter the Dragon" star

86. Like flushed cheeks, colorwise

87. Of reduced degree

DOWN

1. Thick carpets

2. Offshore sight

3. Guanaco kin

4. Madeline of ''Clue''

5. Spot for shots

6. ___ War, 1899-1902

7. Optimistic

8. Egyptian cobra

9. Yon maiden fair

10. He worked on canvases

11. Talkative pet

12. Go through again

13. Like car-radio buttons

14. Acronym

16. Dyeing and marking

21. Have memorized

23. Chinese house idol

27. Common possessive

28. Pub order

29. Cubic meter

31. Rebuilt city north of Cologne

32. Sunday utterance

33. Like an increase from two to twenty

36. Classification system for blood

37. Request payment

38. Adam's rib

39. Part of a basketball hoop

40. Hullabaloo

43. Rifleman's aim improver

44. General address

45. Those elected

46. Of a previous time

48. Contraction in ''The Star-Spangled Banner''

49. Greenish-blue shades

50. '___ Bayou" (1997)

51. Word in a supposed Cagney quote

56. Yunnan or Keemun, e.g.

57. Scale topper, often

58. Mottled horse

59. Unpaid debt

60. Long and impressive

61. In abundance

62. Disease

63. Forgo a hit

66. Reach one's destination

67. Reference or footnote

69. It ran through the veins of Venus

70. 'The Garden of Earthly Delights" painter

72. What the nose knows

74. Prefix for scope

75. It's in the eye of the beholder

76. Not just bad

78. '... ___ nation under God ...''

79. Showed the way

80. Executed

PUZZLE 5

ACROSS

1. Put up, as a picture

5. Blue hue

8. Pharynx and larynx

13. ___ Minor

14. Alternative to acrylics

16. Cast out

17. Drink in a frosted glass

19. Pass on

20. A meeting

22. Discover

23. "Fantasy Island" prop

24. Mythical monster

27. Relating to topology

32. Breathalyzer attachment

35. Way, way off

36. Alpha's opposite

37. "Ah, me!"

39. Affair

42. A homosexual man

43. Infant's illness

45. Part of BYO

47. 40 winks

48. Roman Emperor

51. "I see!"

52. "Act your ___!"

53. "Otherwise..."

58. Hands-on communication?

62. ___ pole

65. Flat grassland

66. Accustom

67. Mine entrance

68. Buzzing pest

69. Fizzle, with "out"

70. Tokyo, formerly

71. Doctrines

DOWN

1. Porter or burden

2. "All kidding ___..."

3. "Teenage Mutant ___ Turtles"

4. Football's ___ Bowl

5. Bar order

6. Hot spot

7. Primordial matter

8. Vertigo

9. Long, long time

10. Come together

11. ___ Today

12. Chester White's home

15. Hex

18. Post-revolution ruling group

21. "The Matrix" hero

24. Arctic ___

25. A subgenre

26. Catch

28. Amiss

29. Amigo

30. People in India

31. 185-country fiscal agcy.

32. ___ Pipes Jig

33. ___ Bowl

34. Australian cockatoo

38. "Dear" one

40. Holiday drink

41. ...

44. A pretender

46. Despot's duration

49. ___ roll

50. Bright circle?

54. Mold and mildew

55. Indian flatbreads

56. Old Irish alphabet

57. Exams

58. Arid

59. Put on board, as cargo

60. All fired up

61. Alliance acronym

62. End

63. "___ moment"

64. "For shame!"

PUZZLE 6

ACROSS

1. Harvest goddess

4. This answer is hard

10. Lowlife

14. Temple, poetically

16. *Motorist's path

17. Strengthen, with "up"

18. A competitor

20. Afflicts

21. Ran through with a cavalry sword

22. Commercial fuel

24. "The Catcher in the ___"

25. Composition

28. Fix, in a way

30. Highlands hillside

31. Gothic architecture features

32. Darkened enclosure

37. Embodiments

38. French door part

39. Smeltery refuse

43. Army unit

45. Manila man

47. Alliance

48. Ticket info, maybe

50. Raising crops

51. A handbook of tables

53. A Muse

57. Gone

58. Toni Morrison's "___ Baby"

59. Lasso or thread cell

62. Calendar abbr.

64. Gluteus muscles

65. "Take that!"

69. Acclivity

70. Province in Canada

74. Affirm

75. A force

76. E-mail, e.g.

77. Bungle, with "up"

78. Incapable of littering

79. ì___ boom bah!î

DOWN

1. Ices

2. Legal prefix

3. High-hatter

4. "Phooey!"

5. Dusk, to Donne

6. Checkers, e.g.

7. "Concentration" pronoun

8. Bust maker, for short

9. 1980's-90's ring champ

10. "Hogan's Heroes" setting

11. metal or wooden

12. Loosen, in a way

13. Complicated situations

15. Peppy

16. At liberty

19. "Don't give up!"

23. Dwarf buffalo of Indonesia

25. Sun, e.g.

26. Coarse file

27. Coffee order

28. "Bolero" composer

29. Adult insect

30. Sarajevo's setting

32. Engine part, for short

33. Battering device

34. Mountain crest

35. Married

36. Archaeological find

39. Shaped spot

40. Jungle climber

41. A deadly sin

42. Attendee

44. Lingerie item

46. Breed

49. Bakery selections

51. Automatic

52. Backstabber

53. Memory trace

54. Experience again

55. Entertains

56. Spuds

60. Desert sight

61. C-___

62. Medicos

63. Addis Ababa's land: Abbr.

65. Fastener

66. ___ milk

67. Doctor Who villainess, with "the"

68. Aims

71. Marienbad, for one

72. E or G, e.g.

73. Absorbed, as a cost

PUZZLE 7

ACROSS

1. Bleat

4. Comply with

8. A lump or mass

12. Duff

13. Anger

14. Black

16. Units of work

17. Carbon compound

18. Eyelashes

19. "... ___ he drove out of sight"

20. Account

21. Darling

23. Store convenience, for short

24. Big sheet

26. Blue

28. Marienbad, for one

30. Electrical unit

32. "Carmina Burana" composer

36. Pronoun in a Hemingway title

39. "Good going!"

41. Narc's find, perhaps

42. "A rat!"

43. Foreword, for short

45. Decline

46. Act the blowhard

48. ...

49. Gross

50. ___ bean

51. It would

52. Anita Brookner's "Hotel du ___"

54. Amniotic ___

56. Arctic

60. "The ___ Daba Honeymoon"

63. Affirmative vote

65. ___ juice (milk)

67. "Go on ..."

68. Kind of cycle

70. Bad end

72. Gull-like bird

73. Antler point

74. Doing nothing

75. Mysterious: Var.

76. Approach

77. Freshman, probably

78. "___ we having fun yet?"

DOWN

1. Grand Canyon transport

2. A rival

3. Balaam's mount

4. Sundae topper, perhaps

5. [Just like that!]

6. "I" problem

7. Canine cry

8. Odd-numbered page

9. ___-Wan Kenobi

10. Pepsi, e.g.

11. Jersey, e.g.

12. Strengthen, with "up"

15. Starchy tuber

20. Chop (off)

22. "A Nightmare on ___ Street"

25. An end to sex?

27. Moo goo gai pan pan

29. "Gimme ___!" (start of an Iowa State cheer)

30. Group of eight

31. Frau's partner

33. Computer architecture

34. ___ jacket

35. Mossback

36. Entanglements

37. "My ___!"

38. "Comme ci, comme

Áa"

40. Acad.

44. Black gold

47. Neon, e.g.

49. "Rocks"

51. Aloof

53. "Give it ___!"

55. Frustration

57. Great Western Forum player

58. Accustom

59. June 6, 1944

60. Matterhorn, e.g.

61. Pat on the back?

62. Small buffalo

64. Cut, maybe

65. Blemish

66. Black cat, maybe

69. Amazon, e.g.

71. "___ to Billie Joe"

72. Caribbean, e.g.

PUZZLE 8

ACROSS

1. Cabinet acronym, once

4. Concrete type

11. Batter's position

16. Absorbed, as a cost

17. Otalgia

18. Kidney enzyme

19. Unsaponified fat

21. ___ out (declined)

22. ___ bread

23. "___ any drop to

drink": Coleridge

24. Daisylike bloom

25. Electromotive force

31. Suppress, in a way

34. Calypso offshoot

35. Born, in bios

36. Bauxite, e.g.

37. Cal. col.

38. Wheedle

41. From now on

44. Always, in verse

45. Santa ___, Calif.

46. Water wheel

48. Component used in making plastics and fertilizer

52. "Arabian Nights" menace

55. Jaded

58. Agree on a contract

62. ___ few rounds

63. Fed. construction overseer

64. "The ___ Daba Honeymoon"

65. Bit

66. Some canines

68. Systems analysis

72. Beats it

73. Beaver's work

74. Beer buy

78. Spikelike inflorescence

79. Party bowlful

83. "Tootsie" Oscar winner

84. Mint family member

85. Pickpocket, in slang

86. ...

87. Make infertile

88. Chester White's home

DOWN

1. Door fastener

2. Decorative case

3. Cried

4. The "p" in m.p.g.

5. Churchill's "so few": Abbr.

6. Victorian, for one

7. Persian attraction

8. Follow, as a tip

9. Dig discovery: Var.

10. "First Blood" director

Kotcheff

11. Gland in men

12. Greek penny

13. Feeler

14. Male hawk

15. Armageddon

20. "___ of Eden"

24. tropical fruit

26. "... or ___!"

27. Book part

28. Sleep on it

29. "It's no ___!"

30. "A rat!"

31. 1922 Physics Nobelist

32. Sundae topper, perhaps

33. Addition column

38. Razor sharpener

39. Blender sound

40. "How ___!"

42. Diamond, e.g.

43. Column crossers

47. Star in Perseus

49. Anger

50. At one time, at one time

51. Indian maid

53. Performers or singers

54. Bucks

56. Motherless

57. Telephone line acronym

58. Consumes

59. Really bad

60. Emerging

61. Infomercials, e.g.

66. Kind of control

67. Carve in stone

69. Argentine dance

70. Dig, so to speak

71. Singers Ruess and Dogg

75. Boosts

76. Fast-moving card game

77. Catch a glimpse of

78. ___ grecque (cooked in olive oil, lemon juice, wine, and herbs, and served cold)

79. Detachable container

80. Burden

81. "___ moment"

82. Atlantic catch

PUZZLE 9

ACROSS

1. The "E" of B.P.O.E.

5. Old Jewish scholars

10. Advil target

14. Wyle of "ER"

15. Extended family

16. Indian Ocean vessel

17. Bounce back, in a way

18. Student getting one-on-one help

19. Drone, e.g.

20. Jet parked on a hill?

23. Bucks

24. Setting for TV's "Newhart"

25. Pasta cut

31. Addis Ababa's land: Abbr.

34. ___ and aahed

35. Gone

36. Alkaline liquid

37. Inside shot?

38. Like an untended garden

40. 100 kurus

41. "___ alive!"

42. "Beg pardon ..."

43. Gillette product

44. ...

45. They may get you off the grid

48. Chop (off)

50. Put up, as a picture

51. The culture

57. Length x width, for a rectangle

58. Fairy tale figure

59. Long, long time

61. Secretary, e.g.

62. An arm and a leg

63. Characteristic carrier

64. Actress Catherine ___-Jones

65. That is, in Latin

66. "... or ___!"

DOWN

1. Charlotte-to-Raleigh dir.

2. Centers of activity

3. Lyricist Gus

4. United States physicist

5. Fifth-century scourge

6. Brown-haired

7. Angler's hope

8. Lying, maybe

9. Escape, in a way

10. Pitcher, of a sort

11. Person or place to another

12. Burrow

13. "Concentration" pronoun

21. Bet

22. Emulated Pinocchio

25. Extremely harmful

26. Blood carrier

27. Shockingly repellent

28. Swelling

29. Certain digital watch face, for short

30. Ballad

32. Austrian province whose capital is Innsbruck

33. Considers judicially

38. First name?

39. Moray, e.g.

40. Contract content

42. Dangerous biters

43. Bluster

46. A line on a sphere

47. Most like a ghost

49. 1970 World's Fair site

51. Certain surgeon's "patient"

52. ___ fruit

53. "Idylls of the King" character

54. Dog command

55. Casting need

56. A long, long time

57. Cutting tool

60. Born, in bios

PUZZLE 10

1	2	3	4			5	6	7	8		9	10	11	12
13					14						15			
16				17							18			
19				20					21	22				
23			24				25	26						
		27			28		29				30	31	32	
33	34	35			36	37				38				
39				40					41					
42				43				44						
45			46				47	48						
		49			50		51			52	53	54		
55	56	57			58	59				60				
61				62	63				64					
65				66				67						
68				69				70						

ACROSS

1. Brewery equipment

5. Fizzy drink

9. Draft

13. Hip bones

14. Graceful bird

15. ...

16. Baked haricot

18. Knowing, as a secret

19. "... ___ he drove out of sight"

20. Malt infusion to make beer

21. Acquire deviously

23. Modus operandi

25. Boot out

27. Priestly garb

29. Mint family member

33. Vortex motion

36. Bat's home

38. Biblical birthright seller

39. Fluff

40. Bona fide

41. "Miss ___ Regrets"

42. "Major" animal

43. Length x width, for a rectangle

44. Con

45. Follow in one's footsteps

47. Marathon

49. Grassy expanse

51. Kind of license

55. Cockeyed

58. "Beg pardon ..."

60. "___ say!"

61. Dynasty in which Confucianism and Taoism emerged

62. In an imminent manner

65. Computer architecture

66. Doesn't ignore

67. Ran, as colors

68. "___ on Down the Road"

69. Astronaut's insignia

70. Waxy part at the base of a bird's bill

DOWN

1. Feelings

2. Having or resembling wings

3. A crude uncouth

4. Ed.'s request

5. Merlin, e.g.

6. Be bombastic

7. Family head

8. Took a phone call

9. A dragon and saved a princess

10. Annex

11. "American ___"

12. Characteristic carrier

14. Big boomer

17. Live

22. Trick taker, often

24. Sauce for seafood

26. Merry

28. Bogeyman

30. ___ Spumante

31. Artless one

32. Boot

33. Aspersion

34. Kind of service

35. Acad.

37. A pint, maybe

40. Leisure time

44. Fairy tale figure

46. Barley bristle

48. Some tournaments

50. Knighted women

52. Deed

53. More out of sorts

54. Glasgow's river

55. "God's Little ___"

56. LaBeouf of 'Holes'

57. Red ink amount

59. "Hey!"

63. "Dilbert" cartoonist Scott Adams has one: Abbr.

64. "ER" network

PUZZLE 11

ACROSS

1. Endure

5. Caper

11. Accused's need

16. "How ___ Mehta Got Kissed, Got Wild, and Got a Life" (Kaavya Viswanathan novel in the news)

17. Park, for one

18. Physics lab device, for short

19. Ore or coal

21. Large lemur

22. Towers over the field

23. "Beetle Bailey" dog

24. Time in power

25. Hate, say

27. Urine

29. Toni Morrison's "___ Baby"

30. A mistake

33. Chocolate and cream

36. #26 of 26

37. Knight stalker

39. "___ bitten, twice shy"

41. Fungal spore sacs

45. West Indies

49. Freshman, probably

50. Assortment

51. Assert without proof

52. Clobber

54. Part of a board

56. Represent falsely

61. When it's broken, that's good

64. Native or inhabitant

65. Call to a foxhound

68. Beachwear

69. Vex, with "at"

72. "Hurray!"

73. City on the Arkansas River

74. A measuring instrument

76. Arab leader

77. Indian turnover

78. "And ___ thou slain the Jabberwock?"

79. Tears

80. Catkins

81. Bad day for Caesar

DOWN

1. Baby

2. Ideals

3. Charades, e.g.

4. "Four Quartets" poet

5. Video maker, for short

6. Of or relating to avionics

7. Blow off steam

8. "Don't bet ___!"

9. Expire

10. Two-year-old sheep

11. Property recipient, at law

12. Hindu phallic

13. Produce a literary work

14. They mix and serve

15. Worst for driving

20. Telekinesis, e.g.

24. Baptism, for one

26. Certain Arab

28. Number of decks

31. Bar

32. Flat braided cordage

34. Not "fer"

35. "Smoking or ___?"

37. "Silent Spring" subject

38. ___ v. Wade

40. 7 zeros; ten million

41. ___-bodied

42. A sleepy person

43. Butt

44. "Rocks"

46. Dried coconut meat

47. Ziti, e.g.

48. Final: Abbr.

52. Begging

53. Sonata, e.g.

55. Loosen, as a cap

56. Concern

57. Bury

58. Hot

59. Finished cleaning

60. Memory trace

62. Quip, part 2

63. Places to sleep

66. Alkaline liquid

67. Police in India

70. Finger, in a way

71. Shakespeare, the Bard of ___

74. Fed. construction overseer

75. More, in Madrid

PUZZLE 12

ACROSS

1. Native or inhabitant

9. In use

13. Coal container

16. Escape clause

17. Dwarf buffalo of Indonesia

18. "Act your ___!"

19. Mottled inlay material

21. Biochemistry abbr.

22. "Gimme ___!" (start of an Iowa State cheer)

23. Bank deposit

24. Astern

25. It may be boring

26. Go over

29. Chat

32. Atlantic City attraction

35. Bar order

36. Assayers' stuff

37. Yams have it

42. "Check this out!"

43. Propel, in a way

44. Dietary laws

45. Cheat, slangily

46. Line of latitude

53. Calendar abbr.

54. Character

55. Romanian money

56. Caribbean and others

58. The quality of not being steady

62. Vintners' vessels

63. Bug

64. Camping gear

65. Fleeting

68. One committing grave crimes?

72. Even if, briefly

73. "To ___ is human ..."

74. ___-bodied

78. Away

79. Anderson's "High ___"

80. Pertaining to a diaphragm

84. "Star Trek" rank: Abbr.

85. A chip, maybe

86. Los Angeles

87. Blonde's secret, maybe

88. Bad look

89. Design or destine

DOWN

1. Church part

2. Who "ever loved you more than I," in song

3. Column style

4. Decide to leave, with "out"

5. Density symbol

6. Aggravate

7. "Not to mention ..."

8. Call for

9. "___, humbug!"

10. Like leftovers

11. A voice exercise

12. W.W. II conference site

13. Outside a barbershop

14. Cause to start burning

15. Most orderly

20. Undertake, with "out"

27. Alias

28. Jelly ingredient

30. Barely get, with "out"

31. Issue

33. Bust maker

34. Sundae topper, perhaps

37. Program that performs repetitive tasks

38. Ring bearer, maybe

39. Hollow wooden

40. Amiss

41. Be silent, in music

45. On the up-and-up?

47. Antiquated

48. Absorbed, as a cost

49. Excellent, in modern slang

50. Harassed

51. ___ gestae

52. Fraternity letters

56. Marked text to keep

57. Sounds

58. Womb-related

59. Recount

60. Marienbad, for one

61. ___ power

66. Cross, maybe

67. "Well, ___-di-dah!"

69. "Come here ___?"

70. Territory-marking stuff

71. Maniacs, slangily

75. Actor Pitt

76. Channel

77. Baker's dozen?

81. The "p" in m.p.g.

82. Big Apple attraction, with "the"

83. "Aladdin" prince

PUZZLE 13

ACROSS

1. British TV network, familiarly (with 'the')

5. Cascades peak

11. According to

16. Bad marks

17. Calling

18. Bob's companion

19. National postal service

22. Human resources head, at times

23. Nth degree

24. Exhausted, with "in"

25. About 1% of the atmosphere

26. Bro, for one

29. Back talk

31. Animal house

32. Innovative projects

36. A head

37. "La BohËme," e.g.

38. Allotment

39. Blue hue

40. Bit

41. Harvest goddess

43. Broke off

44. Psychological attributes

47. Draft horse

50. Grassland

51. Balaam's mount

52. Dash abbr.

55. Lifeless, old-style

56. Banded stone

58. Lady Macbeth, e.g.

59. Landscape or garden

63. Driver's lic. and others

64. "___ Gang"

65. Costa del ___

66. Winged

67. Anniversary, e.g.

69. "My man!"

71. Plain folk

72. A projector

78. Newswoman Shriver

79. Eclipse phenomenon

80. Radial, e.g.

81. Bear

82. Danish money

83. Ponzi scheme, e.g.

DOWN

1. Protestant denom.

2. An ecological activist

3. Position or rank

4. Give to in marriage

5. Cicatrix

6. Consumes

7. "A jealous mistress": Emerson

8. "Comprende?"

9. Course

10. Central Asia

11. Barley bristle

12. Lentil, e.g.

13. A parody

14. Fade away

15. Influence or pressure

20. Children's ___

21. Departure

25. Ancient greetings

26. Septum

27. "Rocks"

28. A kingdom

30. Border guard's demand

33. Study, say

34. Perry Como's "___ Loves Mambo"

35. Ancestry

40. With great care

42. State of the American Union

44. Drove

45. ___ tide

46. ___-friendly

47. A holy thing

48. Asian weaverbirds

49. Genus Monstera

52. Austere and reclusive

53. Excess

54. "Catch!"

57. "Give it ___!"

58. An organism or species

60. 27, to 3

61. Fermented Middle East beverage

62. British title

68. "___ Brockovich"

70. Aroma

71. "When it's ___" (old riddle answer)

73. "We've been ___!"

74. Ace

75. Howard of "Happy Days"

76. "___ moment"

77. "Losing My Religion" rock group

PUZZLE 14

1	2	3		4	5	6	7	8	9	10		11	12	13	14	15
16				17								18				
19			20									21				
22						23				24						
		25	26	27	28			29							30	
31	32	33						34				35				
36				37			38	39			40					
41			42			43				44						
45					46			47			48	49	50	51		
		52	53	54		55				56	57					
58	59	60			61			62				63				
64			65				66			67						
68		69			70	71										
	72				73				74	75	76	77				
78				79			80	81	82							
83				84						85						
86				87						88						

ACROSS

1. ___ de deux

4. Say mean things

11. Pre-euro German money

16. Trick taker, often

17. Mass or tutti of an orchestra

18. "He's ___ nowhere man" (Beatles lyric)

19. Silicon carbide crystals

21. English painter

22. Ticket info, maybe

23. "Dear" one

24. Pago Pago's place

25. Stalls and shows for amusement

31. Multilayer thickness

34. "Catch-22" pilot

35. Batman and Robin, e.g.

36. "Rocky ___"

37. Baby's first word, maybe

38. In a sinuous manner

41. Places for lookouts

44. Automobile sticker fig.

45. Son of Ramses I

46. Attendance counter

48. Ancestry

52. Blah-blah-blah

55. Relating to or concerned

58. Draftee

62. Deception

63. Sun, e.g.

64. "Walking on Thin Ice" singer

65. Newspaper div.

66. Be on one's toes?

68. Common properties of instances

72. Determined container weight

73. Nancy, in Nancy

74. Aquarium

78. Ornamental loop

79. North and South America

83. Organic compound containing the group CONH2

84. Differing

85. "Flying Down to ___"

86. Interpretation or explanation

87. Wired

88. Lizard, old-style

DOWN

1. Agreement

2. Berry touted as a superfood

3. Arid

4. "My man!"

5. Balloon filler

6. PC "brain"

7. Zoologist

8. Obscure

9. Accustom

10. Actor Arnold

11. The unintentional misuse

12. Bouquet

13. Puts differently

14. Sheep of central Asia

15. Arch

20. Highlands hillside

24. Play, in a way

26. Data storage amounts, for short

27. "Once ___ a time..."

28. Astringent fruit

29. "Smoking or ___?"

30. ___ sauce

31. Nervous twitches

32. Bring on

33. Brawl

38. Artificial leg?

39. "Cast Away" setting

40. ___ fruit

42. Cher wear collection

43. Increase, with "up"

47. Acclaim

49. Clickable image

50. Bust maker

51. Dresden's river

53. Grow together

54. Like some humor

56. Conical tent

57. Change

58. Minor player

59. Erstwhile

60. Base

61. "___ say!"

66. Steamed dish

67. The "A" of ABM

69. Chip away at

70. Absurd

71. Like the ZIP code system

75. "God's Little ___"

76. Artless one

77. Bow

78. When it's broken, that's good

79. Quote, part 3

80. A little of this, a little of that

81. Barely get, with "out"

82. Cabernet, e.g.

PUZZLE 15

ACROSS

1. A lever

7. Excellent, to a Brit

13. Ed.'s request

16. Hindu drink of the gods

17. Charge

18. Automobile sticker fig.

19. Between Austria and Switzerland

21. 30-day mo.

22. Deprive of courage

23. Bias

24. Gown fabric

26. "Comprende?"

27. Babysitter's handful

29. Abbr. after a name

31. ___ of Langerhans

32. Trickery

35. A heap

37. Buttonhole, e.g.

39. Newspapers or broadcast media

44. Offer one's couch to, say

46. "Stupid me!"

47. Barbie's beau

48. "Aladdin" prince

49. Broadcasting

50. Cal. col.

52. Join securely

54. ___ grass

55. "___ to Billie Joe"

58. Kills the helper T cells

60. Betelgeuse's constellation

61. Operation of unconscious wishes or conflicts

65. In doubt

66. Dentist's direction

67. Like the Godhead

69. Hot spiced wine punch

72. "Chicago" lyricist

74. "___ alive!"

75. Marienbad, for one

78. Japanese-American

79. ___-Altaic languages

81. Consumed

83. "20/20" network

84. When colder air surrounds a mass

88. Shaggy Scandinavian rug

89. value or esteem

90. City south of Salem

91. PC linkup

92. Glossy fabric

93. One side of a store sign

DOWN

1. Anklebone

2. Nitrogen compound

3. Consider, ponder, or plan

4. 12-point type

5. Italian, e.g.

6. Sylvester, to Tweety

7. Make, as a putt

8. Doofus

9. financial center in northern Belgium

10. Pandowdy, e.g.

11. Action film staple

12. The reappearance in a painting

13. Like some talk

14. Big name in computers

15. "Snowy" bird

20. Saw

25. "Back in the ___"

28. Checkers, e.g.

30. Heavy, durable furniture wood

33. Decorative case

34. War losers, usually

36. "Tarzan" extra

37. Takeoff

38. Kind of cycle

40. Persian potentates

41. Public declaration of intentions

42. Cool

43. Fool

45. Force or extent as to elicit awe

51. Knight fight

53. "___ Brockovich"

56. Kipling's "Gunga ___"

57. "___ on Down the Road"

59. Macho

62. Advocate

63. Distant clouds

64. Barbecue site

68. Beneficial

69. Distort

70. OPEC land

71. An extinct Italic language

73. Thug

76. Pasta choice

77. Chipped in

80. Gulf of ___, off the coast of Yemen

82. Cornstarch brand

85. Bean counter, for short

86. Computer monitor, for short

87. Calendar abbr.

PUZZLE 16

ACROSS

1. Infomercials, e.g.

4. Low energy states

11. Engender

16. "Fancy that!"

17. Kneecap

18. French romance

19. Successive change

21. Long-limbed

22. Computer architecture acronym

23. ___ Solo of "Star Wars"

24. Bear

25. Variety of muskmelon vines

31. Two terms

34. "___ the fields we go"

35. Deception

36. ___-Wan Kenobi

37. ___ roll

38. Interpretable expression

41. Prevent deliberately

44. "The Catcher in the ___"

45. Cut, maybe

46. Corrodes

48. Annexes

52. "Concentration" pronoun

55. Drupe

58. A special group

62. "Catch-22" pilot

63. Victorian, for one

64. "___ we having fun yet?"

65. Batman and Robin, e.g.

66. Super aggressive type, e.g.

68. Ultimate Fighter's favorite side dish?

72. Macbeth, for one

73. Amigo

74. Boosts

78. Ancient Greece's legislative assembly

79. Pitching pro

83. Impulses

84. A baseball drop

85. Grassland

86. Twosomes

87. Busy fellow in a gold rush

88. .0000001 joule

DOWN

1. Biology lab supply

2. Kosher ___

3. Adjusts, as a clock

4. 30-day mo.

5. Masefield play "The Tragedy of ___"

6. In-flight info, for short

7. Alcohol used in antifreeze

8. Homeric epic

9. "Home ___"

10. Gabriel, for one

11. Numerous plants

12. In-box contents

13. Grand Canal sight

14. Improvement in the offspring

15. "Don't give up!"

20. Bounce back, in a way

24. Mideast ruler

26. Black cat, maybe

27. Approaching

28. Avid

29. Moo goo gai pan pan

30. "Comprende?"

31. Be an omen of

32. Bibliographical abbr.

33. Not yet final, at law

38. Ski trail

39. "Beetle Bailey" dog

40. Alarm

42. Any thing

43. Con

47. Condescending one

49. The Everly Brothers, e.g.

50. Acute

51. Antares, for one

53. Great width

54. Liszt's "La Campanella," e.g.

56. "Cogito ___ sum"

57. At liberty

58. Video maker, for short

59. Rhetorical gift

60. Nuts

61. Blouse, e.g.

66. Lead source

67. Boris Godunov, for one

69. ...

70. Milky gems

71. Anklebone

75. "Cast Away" setting

76. Live wire, so to speak

77. Catch

78. Sir, less formally

79. Marienbad, for one

80. Blue hue

81. ___ Dee River

82. "To ___ is human ..."

PUZZLE 17

ACROSS

1. Computer architecture

5. Razor sharpener

10. Urine

14. ___-friendly

15. France's longest river

16. Santa ___, Calif.

17. Beat it

18. Position in a society

20. Surrounds the wheels of a vehicle

22. Absorbed, as a cost

23. "So ___ me!"

24. Babysitter's handful

25. Possessing an education

27. Disney classic

31. Dust remover

32. Counters

33. Baby holder

35. Messy mass

39. High land

40. Do-it-yourselfer's purchase

41. Run off to the chapel

42. Antares, for one

43. Harmony

44. Exterior

45. Beaver's work

47. Unwanted appearance

49. Dissipation of energy

53. Clock number

54. Action film staple

55. Addition

56. Lewis Carroll, e.g

60. Contestant

63. "Mi chiamano Mimi," e.g.

64. Dwarf buffalo of Indonesia

65. Correct, as text

66. Anchovy containers

67. E-mail, e.g.

68. Odd-numbered page

69. Cut, maybe

DOWN

1. Elizabethan collar

2. "Cast Away" setting

3. Caught in the act

4. Credit rating

5. Eat noisily

6. First-rate

7. Cheat, slangily

8. Companion of Artemis

9. Trouble

10. www.yahoo.com, e.g.

11. Certain tribute

12. Any Time

13. Checked out

19. Large intestine

21. In-box contents

25. A rope

26. String together

27. Clappers

28. "Not on ___!" ("No way!")

29. Auto parts giant

30. Gross

34. Allergic reaction

36. French novelist Pierre

37. Airy

38. "Lulu" composer

41. Red dye

43. Industrial plant

46. On the safe side, at sea

48. Cook too long

49. Skywalker Ranch honcho

50. Kind of layer

51. Jake's Women' playwright

52. Gunk

56. Bridge, in Bretagne

57. Bone-dry

58. Peewee

59. "___ of Eden"

61. Crash site?

62. P.I., e.g.

PUZZLE 18

ACROSS

1. Crack, in a way

5. Refuse

10. 100 kurus

14. Maneuvered a ship

15. Accept

16. Carbon compound

17. "O" in old radio lingo

18. Disregard

20. Oater fellow

22. Oolong, for one

23. "It's no ___!"

24. Breed

25. Women only

27. Five Nations tribe

31. Animal house

32. Literally, "dwarf dog"

33. Misfortunes

35. Rear

39. Banded stone

40. Shaggy Scandinavian rug

41. Anatomical dividers

42. Aria, e.g.

43. Brio

44. Santa's reindeer, e.g.

45. Nth degree

47. People are cared for

49. Desk accessory

53. "Bingo!"

54. Control

55. Propel, in a way

56. Spray very finely

60. Socially incorrect

63. Dugout, for one

64. Accommodate

65. Contemptuous look

66. "Empedocles on ___" (Matthew Arnold poem)

67. Coastal raptor

68. Overhangs

69. Beams

DOWN

1. Bonbon, to a Brit

2. Bindle bearer

3. Acknowledge

4. Enjoys seeing sex

5. Polo match

6. Better

7. Mandela's org.

8. Specialty

9. Hit sitcom

10. "Fantasy Island" prop

11. Bring upon oneself

12. Certain tribute

13. Back street

19. Bind

21. Any Platters platter

25. Raise sail or flag

26. Waiting area

27. Wood sorrels

28. Canceled

29. Face-to-face exam

30. Demoiselle

34. Channel

36. "What've you been ___?"

37. Check

38. "Unimaginable as ___ in Heav'n": Milton

41. Group of Bantu languages

43. Stretch

46. Long, long time

48. Don Juans

49. Cursor mover

50. Dog tag datum

51. Deprive of courage

52. "Gladiator" setting

56. On the safe side, at sea

57. Bit

58. Buffoon

59. Flight data, briefly

61. Absorbed, as a cost

62. Gun, as an engine

PUZZLE 19

ACROSS

1. Camp beds

5. Debunk?

10. Teeny finish

16. "God's Little ___"

17. Zambezi River sight

18. Anatomical ring

19. Serving cart

21. Like the State of the Union address

22. Cupid's boss

23. Acquire

24. Fearsome female

25. Not conforming to

29. Bakers' wares

30. Breed

31. "A Prayer for ___ Meany"

32. "Dilbert" cartoonist Scott Adams has one: Abbr.

35. Blue

36. www.yahoo.com, e.g.

37. Opinions or perspectives

39. Bauxite, e.g.

40. "___ Brockovich"

41. "Catch!"

42. Recognized or acknowledged

47. Apple spray

48. Houston university

49. In the capacity of

50. Formerly in love

53. Coke's partner

54. Fed. construction overseer

57. His "4" was retired

58. Centers of activity

59. ___ green

60. "Ah, me!"

61. Calf's head

64. Gooey cake

67. Toni Morrison's "___ Baby"

68. Abreast (of)

69. Heads off

70. Caused by meteorites

72. Soft thin cloth

73. Companion of Artemis

74. Cost of living?

75. Puts in

76. Ski trail

77. Units of work

DOWN

1. Sauce from Tomatos

2. Big Brother's state

3. Attract; cause to be enamored

4. Scraps

5. Density symbol

6. Wildcat, e.g.

7. Satellite connection

8. All in

9. Cracker Jack bonus

10. Rifle attachment

11. Mint family member

12. Necklace item

13. Dermatologist's concern

14. Building additions

15. Beams

20. Amscrayed

26. Shrike's cousin

27. Silver ___ (photography compound)

28. Because of, with "to"

32. Appearance

33. One of TV's Simpsons

34. A pint, maybe

36. Ashes holder

37. At attention

38. Perlman of "Cheers"

39. Creole vegetable

40. Bring out

42. Came down

43. Actors

44. Jalopy

45. Come to

46. Ornamental flower, for short

47. "Much ___ About Nothing"

51. Reassigned from job to job

52. Attention or energy

53. Meditation location

54. Unpleasant opponent

55. Add zest or flavor to

56. Looks

59. Says a whole lot of nothing

60. Shoot for, with "to"

61. Combine

62. Baby berths?

63. Ring bearer, maybe

64. Be slack-jawed

65. Shakespeare, the Bard of ___

66. Camping gear

70. Finish, with "up"

71. "___ to Billie Joe"

PUZZLE 20

ACROSS

1. Doctrines

5. Flexure to the anus

10. Bud

14. Literally, "king"

15. Deflect

16. Bring on

17. Calf-length skirt

18. Excellence

19. Again

20. VC help of the standard kind

23. Amount to make do with

24. Assayers' stuff

25. Stares

28. Hooter

30. Douglas's 'Wall Street' role

34. Slips

36. "___ rang?"

38. Cold and wet

39. Lord Nelson statue site

43. Carbonium, e.g.

44. Short order, for short

45. Fall

46. Finn chronicler

49. "___ alive!"

51. Bits

52. Balkan native

54. "S.O.S.!"

56. 20th-century philosophical movement

62. Language of Lahore

63. Stands for

64. "Beg pardon ..."

66. Unhealthy chest sound

67. ___ of Langerhans

68. Finger, in a way

69. ___ and terminer

70. Autocrats

71. E.P.A. concern

DOWN

1. An end to sex?

2. Gangster's blade

3. Earned

4. Pinkish bell-shaped flowers

5. Part of the mandible

6. Exceedingly

7. Waxy part at the base of a bird's bill

8. Three colors

9. ...

10. Patio furniture

11. Clue

12. Component used in making plastics and fertilizer

13. Cry like a baby

21. Seed covering

22. Hard throw, in baseball

25. "Dig?"

26. Cupid's projectile

27. Hindu teachings on life and force

29. Young's 'Father Knows Best' co-star

31. Dog's coat?

32. Destiny

33. Wide receiver Terrell

35. Priestly garb

37. ___ Today

40. Fluent; easy; superficiality

41. From a bird's feather

42. Social reformer

47. Mint, e.g.

48. After expenses

50. Brawl souvenir

53. Send, as payment

55. Bridge positions

56. 100 cents

57. Inside shot?

58. Doing nothing

59. Western Samoa money

60. Bogus

61. Exec's note

65. Computer storage unit, informally

PUZZLE 21

ACROSS

1. Cold one

5. Disabled

9. Fink

13. A monastery

16. Aquatic plant

17. With trust; in a trusting manner

18. Misses

19. Banana oil, e.g.

20. Bad marks

22. Big Apple attraction, with "the"

23. Foolhardy

25. Absorb

27. Well-known

30. Balaam's mount

32. "Acid"

33. The anus

34. When doubled, a dance

35. Four syllables

38. "Baloney!"

39. Port threatener

41. ___ bit

42. Plunder

44. Lizard, old-style

45. A chip, maybe

46. "___ say!"

47. Car accessory

48. Dukes

49. Gambia's unit of currency

51. Cry

53. In favor of

54. Supplied with an expensive coat

56. Demands

59. Boat in "Jaws"

61. Beautiful people

64. Far from ruddy

65. Five times the speed of sound

66. Beam intensely

67. Greasy

68. ___ a one

DOWN

1. Short order, for short

2. Pink, as a steak

3. Flightless flock

4. Good-for-nothing

5. They get what's coming to them

6. Ashes holder

7. Baltic capital

8. Telephone service

9. Decline

10. Restriction on an activity

11. ...gang aft ___': Burns

12. Strong fiber

14. Lieu

15. Harmony

21. Acclaim

24. Chemistry Nobelist Otto

26. "It's no ___!"

27. Kind of team

28. Dwarf buffalo of Indonesia

29. A family or tribe

31. Cavalry weapon

34. Something to chew

35. Congratulations, of a sort

36. After all deductions

37. Attends

39. Soviet labor camp

40. A white person

43. "Is that ___?"

45. Banking mechanism

47. Large amount

48. Hightails it

49. Backs, anatomically

50. Weariness giveaway

52. Admittance

53. Farm young

55. Conical tent

57. "Two Years Before the Mast" writer

58. Ado

60. Affirmative vote

62. ___ el Amarna, Egypt

63. Aloof

PUZZLE 22

ACROSS

1. "You ___?"

5. Lace-up girdle

11. Kind of mill

16. Arabic for "commander"

17. Kind of recording

18. Extend, in a way

19. Convert code

21. Very, in music

22. Decrease

23. A-line line

24. Arduous

25. It has buckles for a baby

27. Mediterranean plants

29. Oolong, for one

30. Explore

33. Tropical fruit

36. Mandela's org.

37. Aggravating trouble

39. One who takes orders

41. ___ gin fizz

45. A complicated highway

49. Balance parts

50. Carnival attraction

51. Cochise, for one

52. Modern F/X field

54. Bar 'rock'

56. Naturally or artificially impregnated with mineral salts

61. "ER" network

64. Isolated or disjoined

65. A spear

68. Word uttered ecstatically by Homer Simpson

69. Related to the anus

72. Bring up

73. Clearing

74. Grow rapidly

76. Garden tool

77. Annul

78. Above

79. Top competitors, often

80. Speechified

81. Intelligence

DOWN

1. Blue-pencil

2. One-celled protozoans

3. Native or inhabitant

4. Southern breakfast dish

5. "___, humbug!"

6. Ballroom dance

7. Bell the cat

8. Hip bones

9. Swindler

10. ___ roll

11. Written, drawn or engraved

12. Did a double take?

13. Deep down

14. Undercoat

15. Fabrics with diagonal ridges

20. ___ Dee River

24. "Once ___ a time..."

26. Come to mind

28. Big mess

31. Desert bloomers

32. Snakes

34. Units of work

35. U.S. biomedical research agcy.

37. Cooking meas.

38. Marienbad, for one

40. Boot out

41. Attempt

42. Holds a license (degree)

43. "___ Baby Baby"

(Linda Ronstadt hit)

44. Charlotte-to-Raleigh dir.

46. Voice lesson topic

47. Bead material

48. PC "brain"

52. Tycho and Copernicus, for two

53. Goes on and on

55. Thin and slippery

56. Two-winged mosquito-like fly

57. King Mark's bride

58. Minority

59. Sidestepped

60. Sporting individual?

62. Grant

63. Heebie-jeebies

66. Anger

67. It's a wrap

70. ___ Scotia

71. "Thanks ___!"

74. Ace

75. Disobeyed a zoo sign?

PUZZLE 23

1	2	3		4	5	6	7		8	9	10	11		12	13	14
15				16					17				18			
19			20						21							
22						23		24								
	25				26				27				28			
29	30			31			32	33			34	35				
36				37	38					39						
40			41						42							
43		44		45			46	47			48	49	50	51		
		52	53			54	55			56		57				
58	59	60				61					62					
63				64				65								
66			67	68			69	70	71		72					
73		74			75					76			77	78		
79						80		81								
82						83					84					
85			86			87					88					

ACROSS

1. I...

4. All excited

8. Boat with an open hold

12. Bunk

15. "Rocks"

16. Hindu Mr.

17. Voltage through motion

19. Chinese restaurant offering

21. Logician

22. Begin

23. Extreme oldness

25. "___ Gang"

26. Coal carrier

27. Pandowdy, e.g.

28. Absorbed, as a cost

29. Productive period

32. 2004 nominee

34. Celebrations

36. Informal term

39. Incorrect in behavior

40. Branch

41. Climbing bean or pea plant

42. Antiquity, in antiquity

43. Certain protest

45. Blubber

46. "ER" network

48. Boat in "Jaws"

52. 1999 Pulitzer Prize-winning play

54. See-through sheet

57. Long, long time

58. Bud holders?

61. Aggressive

63. Animal hides

64. "Go on ..."

65. Field

66. "___ we having fun yet?"

67. Amigo

69. ___ Today

72. Modern F/X field

73. straight line or lines

76. 1,000 kilograms

79. Power and authority

80. Commits sabotage

82. Security checkpoint

83. Antares, for one

84. Schuss, e.g.

85. Oolong, for one

86. Change

87. Cravings

88. "Comprende?"

DOWN

1. Butts

2. "God's Little ___"

3. ...

4. Eccentric

5. Gangster's gun

6. Witchcraft and sorcery

7. Sea birds; used as fertilizer

8. ...

9. Annoying and unpleasant

10. Certain Arab

11. Inefficient in use of time and effort and materials

12. Narrowly triangular, wider at the apex

13. Clytemnestra's slayer

14. Minimally worded

18. Elephant's weight, maybe

20. Ladies' bag

24. Kind of drive

29. Org. that uses the slogan 'Aim High'

30. Legal prefix

31. Gets rid of

33. Administrative unit of government

35. Edible starchy

37. "I" problem

38. Irish Republic

39. The terminal section of the alimentary canal

44. Blockhead

47. High school class, for short

49. Inclination

50. Chanel of fashion

51. Again

53. Who stimulates and excites people

55. Excessive

56. A muscle which raises any part

58. Most economical

59. Phormio' playwright

60. Euripides drama

62. A homosexual man

64. Parallel or straight

68. "Home ___"

70. Flip, in a way

71. Winged

74. Not just "a"

75. "I, Claudius" role

77. Microwave, slangily

78. Ashtabula's lake

81. Blackout

PUZZLE 24

ACROSS

1. O.K.'s

8. Follower of Mary

12. Little devils

16. Of or relating to a combinatorial system

17. Length x width, for a rectangle

18. Nonexistent

19. Job as a Security Council translator?

21. Cut down

22. "First Blood" director Kotcheff

23. Hawaiian tuber

24. "___ Baby Baby" (Linda Ronstadt hit)

25. His "4" was retired

26. Visual world created by a computer

30. Noble Italian family name

33. Bleat

34. Contradict

35. Court attention-getter

36. "Not to mention ..."

39. Clod chopper

41. Setting for TV's "Newhart"

42. Halfhearted

44. 100 cents

46. Casts

48. Attribute to another source

51. Breastbones

53. Auspices

54. Fits

57. Anita Brookner's "Hotel du ___"

58. "Silent Spring" subject

60. Any thing

62. Cleanse

63. Architectural projection

65. Clinton, e.g.: Abbr.

67. Breakfast choice

69. Arranges personal travel

73. Absorbed, as a cost

74. "... ___ he drove out of sight"

75. "___ of Eden"

76. Fix, in a way

79. Not "fer"

80. Feeds on mistletoe berries

83. Booty

84. "I'm ___ you!"

85. Strong and proud

86. Getaway spots

87. "The Turtle" poet

88. A little slower than moderato

DOWN

1. Adjoin

2. Christmas decoration

3. Like The Citadel, now

4. "A Nightmare on ___ Street"

5. Digestion-related

6. A winged sandal

7. Belt

8. Break out

9. Anatomical ring

10. Hanukkah item

11. Clean

12. Boot part

13. Knowing more than one language

14. Make a plot of

15. Arch

20. "___ rang?"

26. A high official in a Muslim government

27. Walkabout goer

28. Deciduous or evergreen ornamental shrubs

29. Cravings

30. Bon ___

31. #13

32. Decrease in price or value

37. Affranchise

38. "No problem!"

40. "How ___ Has the Banshee Cried" (Thomas Moore poem)

43. Bounce playfully

45. Final notice

47. Language of los Estados Unidos

49. Bit

50. "It's no ___!"

51. Coin opening

52. BÈarnaise ingredient

55. "The Three Faces of ___"

56. Undertake, with "out"

59. Catherine the Great, e.g.

61. Wet

64. Doings

66. Gets rid of

68. Modus operandi

70. Sprite flavor

71. Born, in bios

72. Arum lily

76. Archaeological site

77. "___ It Romantic?"

78. Form of clarified butter

79. "Aladdin" prince

81. fifth (dominant) note of any musical scale

82. Biochemistry abbr.

PUZZLE 25

ACROSS

1. Navajo home

6. Highlanders, e.g.

11. Howard of "Happy Days"

14. Its license plates say "Famous potatoes"

15. Composer Copland

16. Tokyo, formerly

17. Good place for a green thumb

20. Rotten

21. "Silly" birds

22. Amazon, e.g.

24. Word forms having meaning or grammatical function

27. Clothe

28. High-pitched

31. Masefield play "The Tragedy of ___"

33. More, in Madrid

34. Plunder

36. Book of maps

38. Symptoms resulting from neurosis

42. Massage target

43. Representative

45. Ceiling

48. "Aladdin" prince

49. Ziti, e.g.

50. "___ Lang Syne"

52. Sales on television

56. Big ___ Conference

57. Band

59. Breakfast cereal

62. Enlisted personnel

67. "Raiders of the Lost ___"

68. African witchcraft

69. Kitchen gadget

70. ___ de deux

71. Tissue alternative

72. Admittance

DOWN

1. Kills the helper T cells

2. "___ to Billie Joe"

3. comic material for public performers

4. "Beg pardon ..."

5. Do, for example

6. Black

7. Flower part

8. Bauxite, e.g.

9. Deck (out)

10. Catch

11. Cash in

12. "Potemkin" setting

13. Pieces for nine

18. "Is that ___?"

19. Gun, as an engine

22. Branch

23. ___ tide

25. "What's gotten ___ you?"

26. Game ragout

29. Quite a while

30. "The Beast of ___ Flats" (1961 sci-fi bomb)

32. Civil rights org.

35. Indian dish made with stewed legumes

36. Death on the Nile cause, perhaps

37. Ancient colonnade

39. Assortment

40. Behavior often responsive to specific stimuli

41. Commend

44. Jail, slangily

45. 40 winks

46. Dawn goddess

47. A cheap wine of inferior quality

51. Elmer, to Bugs

53. Eskimo boat

54. Aggressive

55. "Absolutely!"

58. "Fiddlesticks!"

60. Fries, maybe

61. A little lamb

63. "Dilbert" cartoonist Scott Adams has one: Abbr.

64. Checkers, e.g.

65. Always, in verse

66. Arid

PUZZLE 26

1	2	3	4	5	6	7		8	9	10	11	12	13	14
15								16						
17								18						
19							20							
21						22					23			24
		25		26					27					
28	29	30					31	32			33			
34				35							36			
37				38						39				
40		41				42								
43				44	45				46		47	48	49	
	50		51					52						
53							54							
55							56							
57							58							

ACROSS

1. Under a curse
8. Breaks out
15. Country house
16. Strong and proud
17. Commercially important food fish
18. Rap variety
19. Restaurant activity
20. Clashed repeatedly with Romans
21. Blueprint details
22. Wrath
23. English informant
25. Sprite flavor
27. Desert bloomers
28. A mortuary
33. "Hee ___"
34. "Walking on Thin Ice" singer
35. Checks for fit
36. "Rocky ___"
37. Head, for short
38. Overwhelming emotion

40. Correct, as text

42. One who goes for the gold?

43. Christmas season

44. Length x width, for a rectangle

46. Chesterfields, e.g.

50. Craft

52. Proenzyme to an active enzyme

53. Provide a bed for animals

54. Compunction

55. Accord

56. Without heat

57. Axes to grind

58. Annoys

DOWN

1. Flu symptoms

2. Bargain-basement

3. Bill of fare

4. Pouch into which the semicircular canals open

5. To insert again

6. Caroled

7. Big jerk

8. "Enigma Variations" composer

9. Squalid

10. N.Y. neighbor

11. Disease of the throat or fauces

12. Ice cream nut

13. Act the host

14. Caribbean, e.g.

20. A lightweight hat worn in tropical countries

22. Thin layers of rock

24. Fuzzy fruits

26. Das Kapital' writer

27. Divided into hundredths

28. Square in Lower Manhattan

29. So lacking in interest

30. A short novel

31. Old Roman port

32. "Oh, my aching head!," e.g.

39. Magnetite, e.g.

41. Put in order

44. Blood carrier

45. Demolishes, in Devon

47. Spreads

48. Money in the bank, say

49. Top competitors, often

51. Curve

52. Atlantic City attraction

53. Grassland

54. Coke's partner

PUZZLE 27

ACROSS

1. Fill in the blank with this word: "Arizona's ___ Mountains"

5. Clone of an optical medium's contents

10. Years on end

14. Veteran journalist ___ Abel

15. Yakked away

16. Iron brace

17. Like Jack Haley in "The Wizard of Oz"?

19. The English translation for the french word: Neva

20. Stand where you lie

21. Underworld leader?

23. Fill in the blank with this word: ""It ___; be not afraid" (words of Jesus): 2 wds."

24. Tampico track transport

25. Golfers' delights

27. any of various elastic materials that resemble rubber (resumes its original shape when a deforming force is removed)

32. Trig angle symbol

33. Zaps, in a way

34. Word of support

35. Unusually excellent

36. Tip

37. Powerful kind of engine

38. WSW's reverse

39. Give ___!' ('Try!')

40. Nintendo dinosaur

41. Unrest

43. Fill in the blank with this word: ""___ pro omnibus, omnes pro uno""

44. Dusseldorf donkey

45. Suffix with social

46. The English translation for the french word: dÈtÈriorer

49. Undoing

54. Whiskey ___

55. Title for Sir Anthony Eden

57. Konrad Adenauer, Der ___

58. Second row

59. Tiny perforation

60. Numerical prefix with oxide

61. Michael who directed the Bond film "The World Is Not Enough"

62. U.S.M.C. recruits: Abbr.

DOWN

1. 1969 Beatles hit

2. Would ___?'

3. Jenny ___ a k a the Swedish Nightingale

4. Museumgoer, e.g

5. Plagiarizes

6. Fill in the blank with this word: ""Young ___ Boone" (short-lived 1970s TV series)"

7. Katherine _____ Porter

8. Stuff

9. Check person

10. Winter 1997-98 newsmaker

11. Upper: Ger.

12. Peaceful race in "Avatar"

13. Women of Andaluc

18. the Dhegiha dialect spoken by the Kansa

22. The English translation for the french word: thÈorie des ensembles

24. Hire

25. Hemp

26. Middle of a famous palindrome

27. The English translation for the french word: nÈcessiter

28. Fill in the blank with this word: ""Hasta ___""

29. Island near Quemoy

30. Parts of masks

31. Renaissance artist Guido ___

32. Something good

36. Noted Carmelite mystic

37. Musical family name

39. Old capital of Romania

40. Jazz's ___ Lateef

42. Ripper

45. Fill in the blank with this word: ""___ you!""

46. Fill in the blank with this word: ""Time ___ a premium""

47. Yard pest

48. Short shot

49. The English translation for the french word: mandat

50. You'll use up 3 vowels playing this word that means toward the side of a ship that's sheltered from the wind

51. Ukrainian city near the Polish border

52. Fill in the blank with this word: "Astronomy's ___ cloud"

53. Fill in the blank with this word: "___ place"

56. The English translation for the french word: appli

PUZZLE 28

ACROSS

1. "___ bitten, twice shy"

5. Chafes

9. Cookbook abbr.

13. Oil source

14. Mideast hot spot

15. King or queen

16. Bride's personal outfit

18. Chilled

19. Bank offering, for short

20. ___ green

21. Back up

23. Starbucks selections

25. Period of the first dinosaurs

27. "Shoo!"

28. Experience

29. Bird ___

30. TV, radio, etc.

33. Mint, e.g.

36. races

38. Certain fir

40. Distort

41. www.yahoo.com, e.g.

42. "Comme ci, comme Áa"

44. "The Open Window" writer

48. Lustrous fabric

51. Movie preview

53. Ambiguous or unclear

54. Hack

55. Caribbean, e.g.

56. Dine at home

57. Stranded at Sugarloaf

60. Certain sorority member

61. Beach bird

62. At liberty

63. New England catches

64. Go to and fro

65. "Bill & ___ Excellent Adventure"

DOWN

1. Ideals

2. Thin

3. The cavity

4. Australian runner

5. Gets promoted

6. Component used in making plastics and fertilizer

7. Bleat

8. Herald, for one

9. One of the Barbary States

10. In seventh heaven

11. Capable of being cut

12. Grand ___ ("Evangeline" setting)

15. Santa ___, Calif.

17. Marienbad, for one

22. Flint is a form of it

24. Sacred songs

25. Mob disperser

26. Blackguard

28. Alpine sight

31. In-flight info, for short

32. Bust

34. "Dear" one

35. Chip dip

36. Doomed

37. "___ calls?"

38. Except

39. Fox relative

43. A limestone

45. Cinch

46. Lamented

47. A decree of a Muslim ruler

49. Moves erratically, as a butterfly

50. "Empedocles on ___" (Matthew Arnold poem)

51. Lion-colored

52. "Chicago" lyricist

54. "The Last of the Mohicans" girl

56. "Yadda, yadda, yadda"

58. "What's ___?"

59. "How ___ Has the Banshee Cried" (Thomas Moore poem)

PUZZLE 29

ACROSS

1. "Bleah!"

4. The "E" of B.P.O.E.

8. Intensifies, with "up"

12. Beaver's work

15. Ed.'s request

16. N.Y. neighbor

17. Narcotic drug

19. Kind of soup

21. Excerpts from a literary work

22. Copy

23. Landlocked country in central Asia

25. "Spy vs. Spy" magazine

26. Gloomy

27. ...

28. "___ we having fun yet?"

29. Emulates Sarah Hughes

32. ___ system

34. Hotel posting

36. Marry within the same ethnic

39. Least tainted

40. "Them"

41. Moving

42. Biochemistry abbr.

43. Cookbook abbr.

45. Mary ___ cosmetics

46. High school class, for short

48. "Ali ___ and the 40 Thieves"

52. "It's no ___!"

54. Tom, Dick or Harry

57. Masefield play "The Tragedy of ___"

58. Ready for shipping

61. Captain's underling

63. Greek penny

64. Balaam's mount

65. End of a threat

66. A pint, maybe

67. Delay

69. Astern

72. Brown, e.g.

73. Less serious than a felony

76. Crows' homes

79. Show a certain behaviour

80. A degree of dominion

82. Show

83. Caribbean and others

84. "The Matrix" hero

85. "Baloney!"

86. Pasturelands

87. Dangerous biters

88. "Silent Spring" subject

DOWN

1. "Back in the ___"

2. Be slack-jawed

3. A helpful partner

4. The outer germ

5. Chop (off)

6. Work, as dough

7. Big mess

8. Asian nurse

9. Where the Grimaldis reign

10. Life force in yoga

11. Articles or advertisements changed

12. Give orders

13. Star in Scorpius

14. Most malicious

18. "For ___ a jolly ..."

20. About to explode

24. Distort

29. Free of coarseness

30. Door feature

31. Preserved, in a way

33. "So long"

35. Kuwaiti, e.g.

37. "The ___ Daba Honeymoon"

38. Saudi Arabian bucks

39. "Now!"

44. Short shot

47. Chit

49. A person undergoing psychoanalysis

50. Embargoes

51. A chip, maybe

53. It rises every year

55. Month after Adar

56. Creepy feeling

58. Climb awkwardly, as if by scrambling

59. Type of sculpture (Var.)

60. (slang) very angry

62. "Well, I ___!"

64. Government group

68. Nitrogen compound

70. Bone cavity

71. Sets right

74. "Dear old" guy

75. Long, long time

77. 20-20, e.g.

78. Mucus

81. Blast

PUZZLE 30

ACROSS

1. "Cast Away" setting

5. Heroin, slangily

9. Consume

14. A tiny amount

16. African antelope

17. Resembling a Mongol

18. ___-off coupon

19. Most strict

20. The beach at a seaside resort

21. "Green Gables" girl

22. A printer's hand-inking cylinder

23. Fix firmly

25. Keepsake

27. Pool contents?

28. Call

29. Bill and ___

30. "Back in the ___"

31. Come to mind

32. Copter's forerunner

33. Auction offering

34. Chin indentation

35. Genuflected

36. Tastelessly showy

38. Works with feet?

39. Stress results, maybe

40. "Frasier" actress Gilpin

41. Maximum

42. Drummer

46. Encumbrances

47. Dismiss

48. Present

49. A salt formed by the union of arsenious

50. Chips in

51. One's final stand?

52. Copper

DOWN

1. Doctrines

2. Sean Connery, for one

3. Ancestry

4. Hogarth, for one

5. Mum

6. Intimate

7. Came down

8. "Crikey!"

9. Bounded area

10. Long bony fingers

11. The state of being tangent

12. In an ulterior manner

13. Hail Mary, e.g.

15. Metric heavyweights

20. Cut back

22. Hike

23. Denouement

24. A payment of part of a debt

25. Z's neighbor

26. Carrot, e.g.

27. Split

28. Iron

31. Argus-eyed

32. Relating to, or resembling, gneiss

34. A board game

35. Like Hyundai and Samsung

37. Light fixture

38. A heavy rain

40. Antiquated

41. Arm bone

42. ...

43. Bit

44. Gulf of ___, off the coast of Yemen

45. Home, informally

47. Hack

PUZZLE 31

ACROSS

1. Loudly commends

6. Dismissive exclamation

10. Fill in the blank with this word: "___, Crackle and Pop"

14. Fill in the blank with this word: "Caput ___ syndrome (arm problem)"

15. Green-skinned dancing girl in a "Star Wars" film

16. Must've been something ___'

17. Attractive bar

20. GM: "___ the USA in your Chevrolet"

21. Stan on the sax

22. Opposite of 46-Across

23. Quicken Loans Arena cagers

24. Love __' (Beatles song)

26. See 20-Across

32. Quadrennial polit. event

33. Chemistry Nobelist Otto

34. Zoologist's foot

35. Pie ___ mode

36. Seine-___, department bordering Paris

39. Kellogg's Cracklin' ___ Bran

40. What ___ told you ...?'

41. Sony subsidiary

42. Mein ___

43. Show contempt for yellow fruit?

48. Swiss snowfield

49. Uproars

50. 1994 sci-fi epic

53. New York's ___ Island

54. Therapists' org.

57. Refrain from eating pasta?

61. Sham

62. Launch ___

63. Weird

64. Pack ___ (quit)

65. The Beatles' "Back in the ___"

66. St. Gregory's residence

DOWN

1. Cadence syllables

2. You'll use up 3 vowels playing this word that means toward the side of a ship that's sheltered from the wind

3. Words often before a colon

4. On the ___ (fleeing)

5. Under Siege' star

6. Westminster Abbey area

7. TV character who jumped the shark, with "the"

8. Last: Abbr.

9. Fill in the blank with this word: ""___ Woman" (1972 #1 song)"

10. Turin title

11. -

12. To ___ (unerringly)

13. Czech-born N.H.L.'er Sykora or Prucha

18. The English translation for the french word: Neva

19. "Homage to Clio" poet and family

23. Year St. Pius I died

24. Unscramble this word: neam

25. Textbook market shorthand

26. Stout detective Nero ___

27. We build castles ___ when flushed with wine and conquest': Butler

28. Finish this popular saying: "Talk is_____."

29. Fill in the blank with this word: ""You're ___ and don't even know it""

30. William Blake: "When the stars threw down their spears,/ And watered heaven with their _____"

31. Old Testament book

32. The English translation for the french word: caÔd

36. Actress Frost and others

37. Top-___ (leading)

38. Fill in the blank with this word: "Allan ___, "Sands of Iwo Jima" director"

42. Heave-___ (dismissals)

44. Kind of statement, to a programmer

45. Unscramble this word: thrrea

46. Family name suffix in taxonomy

47. Joseph of "Citizen Kane"

50. Religious mystic

51. Steps down to a river, in India

52. Japanese golfer Isao ___

53. Three-stripers: Abbr.

54. They're what a pompous person "puts on"

55. Petits ___ (French peas)

56. Ralph Vaughan Williams's "___ Symphony"

58. The Nittany Lions: Abbr.

59. Sue Grafton's "___ for Alibi"

60. Writer Josephine

PUZZLE 32

1	2	3	4		5	6	7	8	9		10	11	12
13					14						15		
16				17							18		
19					20					21	22		
			23					24					
25	26	27					28				29	30	31
32						33				34			
35					36					37			
38					39					40			
41				42					43	44			
			45					46					
47	48	49					50				51	52	53
54					55	56					57		
58					59						60		
61					62						63		

ACROSS

1. something acquired without compensation

5. "Love Story" author Segal

10. South Africa's ___ Paul Kruger

13. Giant chemicals corporation

14. Old coin worth five centesimi

15. "___ Angel" (Mae West film)

16. Steal a pass?

18. Change

19. Bao ___ (former Vietnamese emperor)

20. Old English letters

21. More lean and muscular

23. "Trick" joint

24. Citrus fruit

25. Stingers?

28. Open swath in a forest

32. informal or slang terms for mentally irregular

33. Double contraction

34. Chemical endings

35. It doesn't stand up straight: Abbr.

36. Cager Gilmore

37. Things sometimes exchanged

38. Vincent Lopez's theme song

39. Russian auto make

40. Autobahn auto

41. Nuts, raisins, dried fruit, etc.

43. a town in north central Louisiana

45. Call

46. Science fiction author A. E. van ___

47. Scrappy fellow?

50. Vamp Theda

51. TV drama set in Las Vegas

54. "The Alexandria Quartet" finale

55. Occasions to cry "Eureka!"

58. Abbr. on an envelope

59. Williams title start

60. Kosher ___

61. an agency in the Department of Transportation that is responsible for the safety of civilian aviation

62. Exceptional rating

63. Bingo call

DOWN

1. made from or covered with gold

2. Pelvic parts

3. Poodle name

4. explosive consisting of a yellow crystalline compound that is a flammable toxic derivative of toluene

5. Stop sign feature

6. Valium manufacturer

7. Varieties

8. Do, re, mi

9. Starts illegally?

10. Sacred Buddhist mountain

11. French wave

12. open land usually with peaty soil covered with heather and bracken and moss

15. having no bearing on or connection with the subject at issue

17. Certain race ... or a cryptic title to this puzzle

22. The Beatles' "___ Mine"

23. Pitcher who says "Oh, yeaahh!"

24. "The Sopranos" matriarch

25. Singer Black

26. rotating mechanism consisting of an assembly of rotating airfoils

27. City due west of Daytona Beach

28. Cremona violin

29. "Peace ___ time"

30. Change at the top?

31. "Ah, Wilderness!" mother

33. Train track beam

36. Popular references

42. ___ Fail (Irish coronation stone)

43. Ninnies

44. Old Irish alphabet: Var.

46. "Let's go, Miguel!"

47. Military org. with the motto "Per ardua ad astra"

48. ___ Loma, Calif.

49. "___ life!"

50. informal or slang terms for mentally irregular

51. Online tech news resource

52. Surprise

53. "Beauty ___ the eye Ö"

56. Consumes

57. Teacher's deg.

PUZZLE 33

ACROSS

1. Plasm prefix

5. Variety has long used this word for a box office hit

10. Nonsense

14. The English translation for the french word: naÔf

15. Two strikes?

16. Tiers ___ (French commons)

17. Tournament passes

18. Part of a metropolitan region

19. Gillette ___ Plus

20. tough elastic tissue

22. Radical

23. Inc. cousin

24. Flower parts

26. Unite

30. Winner of all four grand slam titles

32. Fill in the blank with this word: "___ del Fuego"

34. Cries of pain

35. Sgts., e.g.

39. Fit of shivering, in dialect

40. Early statistical software

42. Zipped

43. Zaire's Mobutu ___ Seko

44. Yahoo! competitor

45. Phrase of irresolution

47. Shipping weight

50. Fill in the blank with this word: "Correo ___"

51. Sewing groups

54. Son of, in Arabic names

56. Fill in the blank with this word: "Allegro ___ (very fast)"

57. Viniculturist's sampling tube

63. Fill in the blank with this word: "Dragon's ___ (early video game)"

64. Holy, to Horace

65. Mae West's '___ Angel'

66. Volunteer org. launched in 1980

67. When lunch ends, maybe

68. Mineral residue

69. Word with code or road

70. Loosens (up)

71. With 52-Across, what angels pray for

DOWN

1. You might take investing tips from this network's "On the Money" or "The Call"; Jon Stewart probably doesn't

2. Fill in the blank with this word: ""Divine Secrets of the ___ Sisterhood""

3. Top-___ (leading)

4. Fill in the blank with this word: ""The Bells ___ Mary's""

5. Lymphocyte found in marrow

6. State of southern Mexico

7. Lively '60s dance

8. Raccoon's hands

9. White House fiscal grp.

10. With 59-Across, indication of caring

11. Holy Roman emperor, 962-73

12. Pope John Paul II's real first name

13. The Louvre's Salles des ___

21. The Sopranos' actor Robert

22. Worrying sound to a balloonist

25. Japanese immigrant

26. Chairmen often call them: Abbr.

27. Yeats's land

28. Defendant at law

29. Work necessities, for some

31. With 70-Down, do much (for)

33. It's ___!' ('We'll go out together!')

36. Rope fiber

37. Writer Sarah ___ Jewett

38. Zaire's Mobuto Sese ___

41. Neighbor of South Africa

46. Fill in the blank with this word: ""___ pis!" ("Too bad!," in France)"

48. Ransom ___ Olds

49. Prime-time time

51. Serenity, in Seville

52. Writer Asimov

53. Where to sign a credit card, e.g.

55. Earthen embankments

58. Writer of sweet words?

59. Bar sounds

60. The English translation for the french word: imam

61. Requests for developers: Abbr.

62. Below-ground sanctuary

64. You reap what you ___'

PUZZLE 34

ACROSS

1. Seeds lawns

8. Helping hand for the Addams family?

13. Do a museum job

14. Ancient Tuscany

16. Decorate or cover lavishly (as with gems)

17. A conversation between two persons

18. Castle protector

19. Enter†into a list of prospective jurors

20. Send out

21. Grafton's '___ for Lawless'

22. Nest egg, for some

23. Fix code

25. Downright dirty

27. Virtual person in a computer game

28. Small oval seeds of the sesame plant

31. Draw on

32. Tending toward the political left

33. Employ

36. A person who attests to the genuineness of a document or signature by adding their own signature

40. Chicken ___

41. Walruses lack them

42. Completely destroy

44. Best

45. Wheeler-dealer, for short

46. Blood pigment

47. Before birth

49. Rome's 'Palace of the Popes'

52. Albino's lack

53. A kind†of heavy jacket

54. Soul

55. Dugong cousin

56. Stone marker

57. Craft

DOWN

1. Bellyached

2. Gives a new title to

3. Attribute

4. Low, extensive cloud

5. Stock offering?

6. Limerick language

7. Rectangular paver

8. Course hazards

9. Polynesian performance

10. O. Henry, for one

11. Lagos locale

12. Old streetlight

14. "Sound of Music" tune

15. Showing fear and lack of confidence

24. Wingding

25. Refrain from changing, disturbing or taking

26. Banana oil, for one

29. Markedly rapid

30. Indian stewed legume dish (Var.)

33. A period†of†time when something (as a machine or factory) is functioning and available for use

34. Next

35. To drive†out; to expel

37. Khatami's capital

38. Needed Pepto-Bismol, perhaps

39. English-language versions, maybe

43. Shutterbug's accessories

45. Trattoria choice

48. Chinese unit of weight

49. Hightails it

50. Uptight

51. Chic

PUZZLE 35

ACROSS

1. Fill in the blank with this word: ""Give ___ to Cerberus" (Greek and Roman saying)"

5. Publicly exaggerate, in slang

10. ___ punk (hybrid music genre)

13. Spanish skating figures

15. What online shoppers may spend

16. N.F.L. Hall-of-Famer ___ Barney

17. Feast of Trumpets

19. Fill in the blank with this word: ""___ Blue?" (1929 #1 hit)"

20. Chorus parts

21. Sent regrets, say

23. Fill in the blank with this word: "At ___ rate"

24. Fill in the blank with this word: "___ tai"

25. Recording session need

27. Fateful event for the Titanic

31. Fill in the blank with this word: "___-American"

34. Zebulun's mother, in the Bible

35. Night, to Nero

36. Boxing sobriquet

37. 'Vette rival

39. Fill in the blanks with these two words: "____ time"

40. Fill in the blank with this word: "Double ____"

41. Heater stats

42. Leveled, in London

43. Spaghetti-in-a-can icon

47. Sporty car features

48. Fill in the blank with this word: "____ in apple"

49. Treebeard, e.g.

52. Wax-wrapped product

54. Dust-laden winds

56. Fill in the blank with this word: "____-hoo"

57. Reason to call the exterminator?

60. Piece of the rock' company, informally

61. Suffix with hippo-

62. Poet William Rose ____

63. Ways around: Abbr.

64. Sides of some ancient temples

65. Sound of a leak

DOWN

1. Vital carrier

2. The English translation for the french word: mÈpriser

3. Key's opener?

4. Sci-fi writer Frederik

5. One of two sides of a story?

6. Worms cries

7. Plaintive cry

8. Swabs' grp.

9. Villain in Exodus

10. Colonial captives

11. Dole's running mate, 1996

12. Surrounded by

14. Tribal V.I.P.

18. Jai ____

22. Word repeated in Mad magazine's "____ vs. ____"

25. Toothpaste ingredient

26. Spot on ABC

27. World view

28. Fill in the blank with this word: ""I Still See ____" ("Paint Your Wagon" tune)"

29. Used up

30. Gen. follower

31. Fill in the blank with this word: ""____ Live," 1992 multiplatinum album"

32. Fill in the blank with this word: "____ Tzu (toy dog)"

33. Trigger-happy, say

37. 'Vette option

38. Pays the price for

39. Popular cable channel

41. Rustic carriages

42. Soak up again, as liquid

44. Teleflora competitor

45. Montreal Expos legend Tim

46. Ste. Jeanne ____

49. Redeemable recyclables

50. What counters may count

51. Need for the winner of a Wimbledon men's match

52. Swindles

53. Fill in the blank with this word: "Astronomy's ____ cloud"

54. rock that form the continuous lower layer of the earth's crust

55. Unexciting marks

58. Fill in the blank with this word: ""If the ____ is concealed, it succeeds": Ovid"

59. Fill in the blank with this word: ""____ bad!""

PUZZLE 36

ACROSS

1. Chocolate-covered caramels from Hershey

6. Shucks!'

10. Klinger player on "M*A*S*H"

14. Geo model

15. Fill in the blank with this word: ""Parlez-___ fran"

16. Greek goddess Athena ___

17. making a false outward show

19. What an A is not

20. Like rosebushes

21. Tiny criticism

22. Unscramble this word: vegi

23. Fill in the blank with this word: "Basketball's ___ Elmore"

25. Lion tormentors

27. Scratch at the door?

32. She, in S

33. Local theater, slangily

34. Classic spy plane

36. Warwick's "___ Little Prayer"

40. Wreck-checking org.

41. Pot creators

43. "King Lear" or "Hamlet": Abbr.

44. Where Jean-Claude Killy practiced

46. Fill in the blank with this word: "Andean peak ___ Cruces"

47. When Othello decides

he wants to poison Desdemona, this villain suggests that he strangle her instead

48. Romanian money

50. Madonna role, 1996: 2 wds.

52. Repeated word in a contract

56. Siglo de ___ (epoch of Cervantes)

57. Fill in the blank with this word: "Days of ___"

58. Onetime TWA rival

60. String substitute?

65. Money writer Marshall ___

66. Really mad

68. Fill in the blank with this word: ""___ put our heads together ...""

69. Fill in the blank with this word: "___ podrida"

70. Tipsy

71. Dusseldorf donkey

72. Fill in the blank with this word: "___ fixe"

73. Fill in the blank with this word: "___ voce"

DOWN

1. Tach readings

2. The English translation for the french word: Oran

3. Fill in the blank with this word: ""Mona ___""

4. Fill in the blank with this word: "Broadway's ___ Simon Theatre"

5. in a smug manner

6. TiVo, for one, in brief

7. Stable color

8. Thorough check

9. Fill in the blank with this word: "___ fly"

10. What it's like to be Spider-Man?

11. When ___ said and done

12. Town council president, in Canada

13. Hard-to-find cards, to collectors

18. Hottie

24. Wrapper stat.

26. Ransom ___ Olds

27. Sicilian resort city

28. One of the major leagues: Abbr.

29. Shortening used in recipes

30. 1974 David Bowie song

31. They hold water

35. New York city near Binghamton

37. William Saroyan's "My Name Is ___"

38. Fill in the blank with this word: "___ Sant'Gria

(wine)"

39. Fill in the blank with this word: ""A Girl, A Guy and ___" (1941 Ball movie)"

42. This fish-eating bird of prey is also known as the fish hawk

45. GM: "___ the USA in your Chevrolet"

49. Classic 1896 Alfred Jarry play

51. Words to swear by

52. Poet Elinor

53. Goes by foot, with "it"

54. Fill in the blank with this word: ""Maid of Athens, ___ part": Byron"

55. Word on a three-sided sign

59. Sound heard through a stethoscope

61. Sony founder Morita

62. Vocal rise and fall

63. Frankie Laine's "___ Her Go"

64. Fill in the blank with this word: "Actor ___ Ray"

67. Peace Nobelist Kim ___ Jung

PUZZLE 37

ACROSS

1. Train stop

6. Fill in the blank with this word: "Fashion's ___ Saint Laurent"

10. Peek-___

14. Declarer

15. The English translation for the french word: groin

16. Fall mos.

17. Fill in the blank with this word: ""That ___ lady ...""

18. On bended ___

19. Fill in the blank with this word: ""___ Death" (Grieg work)"

20. It's classified

23. Sorority letters

25. Withheld

26. Fill in the blank with this word: ""___ sorry!""

27. Stuck

29. Key of Schubert's Symphony No. 5

32. Visit

33. South American monkey

34. Some undergrad degs

37. Features found in 17- and 64-Across and 11- and 28-Down

41. Fill in the blank with this word: "___ out a win"

42. Explorer John and others

43. Play co-authored by Mark Twain

44. Made off with

46. Provincial capital in NW Spain

47. Year's record

50. Fill in the blank with this word: "___ sponte (of its own accord, at law)"

51. Tight end, at times

52. Concerns of Archimedes

57. See 70-Across

58. Word-of-mouth

59. Parts of masks

62. Potato source

63. Mother and wife of Uranus: Var.

64. Hawaiian feasts

65. Ward on TV

66. Three-stripers: Abbr.

67. Raptor 350 and others

DOWN

1. Thomas Bailey Aldrich story "Marjorie ___"

2. Rice and Lloyd Webber's "Waltz for ___ and Che"

3. Quickly

4. Wine: Prefix

5. Den decorations

6. On the team?

7. Price of a movie?

8. D.O.E. part: Abbr.

9. Fill in the blank with this word: "___-Ball"

10. Fill in the blank with this word: ""Vigilant ___ to steal cream": Falstaff"

11. Twiggy broom

12. Subjective pieces

13. Bone: Prefix

21. Fill in the blank with this word: "___-string"

22. Fill in the blank with this word: "___ Maria"

23. Watchmaker's unit of thickness

24. Thumbing-the-nose gesture

28. Late ___

29. Snake, for one

30. Quarters

31. Finish this popular saying: "Let sleeping dogs_____."

33. Purveyor of nonstick cookware

34. Nine

35. Fill in the blank with this word: ""___ always say ...""

36. What to call un hombre

38. Tulsa sch. named for an evangelist

39. How bitter enemies attack

40. Fill in the blank with this word: ""The Sweetheart of Sigma ___""

44. Violinist Heifetz

45. She, in S

46. Arles assent

47. Sap-sucker's genus

48. [See title, and proceed]

49. Mass communication?

50. Ward and namesakes

53. Yuletide quaffs

54. Zip strip?

55. Fill in the blank with this word: ""Hurry up and ___""

56. Sign of a hit show

60. Mao's successor as Chinese Communist leader

61. Worrying sound to a balloonist

PUZZLE 38

ACROSS

1. I.R.S. form 1099-___

5. Proceeded slowly

11. Fill in the blank with this word: "#1 ___ (gift T-shirt slogan)"

14. Welcome, as a visitor / Try to make a date with

15. Response to 'Gracias'

16. Spanish queen until 1931

17. Come down hard

19. The English translation for the french word: lÈpisostÈe

20. Researcher's wear

21. Shipmate of Starbuck

23. Motel freebie

24. Noah of 'Falling Skies'

26. Mulling

27. Convocation of witches

29. Role in "Son of Frankenstein"

32. Fill in the blank with this word: "___ Nordegren, ex-wife of Tiger Woods"

33. Fill in the blank with this word: "___ sister"

35. Run of letters

37. Trail

38. Writer-turned-physician?

41. What's funded by FICA, for short

43. Swimming great Diana

44. With 17-Down, a temporary urban home

45. Plains dwelling: Var

47. What 17-, 25-, 37-, 52- and 62-Across are, themewise

49. Raptor 350 and others

53. Support

54. Tennis miss

56. Old Dodge

57. Designer with the Rock Me! fragrance

61. Went wide of

63. The English translation for the french word: journaliser

64. Old-time journalist-socialite

66. Bygone daily MTV series, informally

67. Leader of the pack

68. Would ___?'

69. What a farmer bales

70. Brother comic Shawn or Marlon

71. Scottish rejections

DOWN

1. Oscar-winner Matlin

2. Susan who wrote the best seller "Compromising Positions"

3. Winter weather wear with adjustable straps

4. Pres., to the military

5. Start of an opinion

6. With precision

7. This is ___'

8. Fill in the blank with this word: ""___ known then.""

9. Byrnes and Hall

10. Word repeated in "Now ___ away! ___ away! ___ away ...!"

11. European air hub

12. Haemoglobin deficiency

13. The English translation for the french word: chÈrie

18. Western burg, unflatteringly

22. ___ Field

25. Pushed

28. Venomous snake

30. Cry from one who just got the joke

31. Jamaica's Ocho ___

34. Hopalong Cassidy portrayer

36. Peer group setting?

38. With the mouth wide open

39. Genealogist's abbr.

40. Fill in the blank with this word: ""___ wise guy, eh?""

41. Spy aircraft's attribute

42. Madame in Roma

46. Rose ___ rose..'

48. The English translation for the french word: quartier-maÓtre de deuxiËme classe

50. Ford from long ago

51. Title girl in a 2001 French comedy

52. Moves laterally

55. 30-second spots, e.g.

58. Brine

59. "The Producers" role

60. TV show for which Bill Cosby won three Emmys

62. Heads ___, tails..'

65. Fill in the blank with this word: "___ moment"

PUZZLE 39

1	2	3	4	5		6	7	8	9		10	11	12	13
14						15					16			
17				18							19			
20						21					22			
			23		24				25					
26	27	28	29		30			31						
32			33	34			35				36	37	38	
39								40						
41					42		43							
		44			45				46					
47	48	49			50			51	52					
53				54				55		56	57	58	59	
60				61		62								
63				64				65						
66				67				68						

ACROSS

1. Cookie amount

6. Highest man or lowest woman

10. Make a sweater, perhaps

14. Be-Bop-___' (1956 hit)

15. Duds

16. Dollar prefix

17. Extremely luminous

19. Black, in verse

20. Political escapee

21. Hardly gregarious

22. Cod cousin

23. It creates an uplifting baking experience

25. Capital of Tibet

26. Keeps company with

30. Dread

32. Slaughtered and dressed for food

35. Force or drive back

39. Place into groups

40. Covered with, or having many

41. Overhang in a sports bar?

43. Spaghetti specification

44. Ball in cricket

46. One who may make an admission?

47. Z's neighbor

50. Life preserver material

53. Balsam

54. Track record?

55. One who creates a scene?

60. Seed coating

61. Employee of a railroad

63. Native of France

64. Indigenous Japanese people

65. Family member

66. Abounding with elms

67. Industrial curse

68. Spread a little joy, e.g.

DOWN

1. Part of a diamond?

2. One with an old school tie?

3. South American Indian people

4. sucks blood

5. He met Sally

6. Of a previous time

7. Luxuriant

8. Modern replacement for 108-Down

9. One-time marriage vow word

10. To haul under the keel of a ship

11. Woolen scarf

12. Shackles

13. South Pacific nation

18. Genealogy word

24. Chest thumper

25. Fatty compound

26. Beginning of healing, sometimes

27. Effortlessness

28. It may be heard in the Highlands

29. Without respect

31. Boxer's cue

33. Specialized vocabulary

34. Ticket part

36. Waiter's place

37. Words with tone or poor example

38. Famous apple site

42. Prepare to fire

43. High mountain

45. Hispanic American

47. Put on, as a production

48. Occurring every 60 minutes

49. Ancient Troy

51. Stroke's need

52. Money of Denmark

54. Arm, to Armand

56. Deer's scut

57. Inventor's ''light bulb''

58. Religious faction

59. Branch headquarters?

62. Type of nut

PUZZLE 40

1	2	3	4	5		6	7	8	9		10	11	12	13
14						15					16			
17					18						19			
20					21				22		23			
			24					25						
	26	27				28					29	30		
31						32					33		34	
35					36					37	38			
39				40			41							
	42		43	44			45							
		46				47								
48	49	50					51				52	53	54	
55				56		57				58				
59				60					61					
62				63					64					

ACROSS

1. Distinctive style

6. Turns on the waterworks

10. Word to a fly

14. Gold unit

15. Healthy berry

16. Condor's pad (Var.)

17. Surface layer

19. Unfettered

20. Archer's wood

21. To boot

22. Beat a dead horse

24. Farm baby

25. Kuwaiti prince

26. Piece†of†work

31. Doohickey

32. Chaucerian tale-teller

33. 'Ay, there's the ___"

35. More green around the gills

36. 'Dead man's hand" card

37. Coward's ''To Step ___"

39. 'My ___" (Mary Wells classic)

40. Jamaican pop music

41. Wince

42. likeness or counterpart

46. Moonshine ingredient

47. Bit of nuclear physics

48. He gets what's coming to him

51. One-customer connector

52. Baseball great Mel

55. Something fishy?

56. Evergreen perennial

59. Gal's sweetheart

60. Personal appearance

61. Celery

62. Wraps up

63. Ratted

64. Disintegrates, as a cell

DOWN

1. Fuddy-duddy

2. Inveigle

3. Yet again

4. Cooler cooler

5. A rented car

6. Prune

7. 'L'___ del Cairo" (Mozart opera)

8. Grill or fire

9. Jazz band instrumentalist

10. Hunting journey, often

11. Basil or sage, e.g.

12. Two-colored cookie

13. ___ and terminer

18. Like a rock

23. Torched

24. Relinquish control

26. Alone

27. Tentacled creature

28. Machu Picchu denizen

29. NASA problem part

30. Urge physically, but gently

31. Archaeological excavation

34. One who's always buzzing off?

36. Motionlessness

37. Heraldry or heraldic

38. Famous twins' birthplace

40. Mountain waterways

41. Quoted as an authority

43. His wings melted in the sun

44. Little tyke

45. Wonder or awe

48. Solitaire spot, perhaps

49. Famous apple site

50. Prompt

52. Singer Redding

53. Undeniable

54. Highland hats

57. It's mightier than the sword

58. Finder of secrets

PUZZLE 41

ACROSS

1. Ball field covering

5. "Check this out!"

9. High-hatter

13. Aroma

14. Blood of the gods

16. Bring on

17. Anniversary, e.g.

18. Nigerian monetary unit

19. ___-friendly

20. Harsh Athenian lawgiver

22. Position in a graded series

24. Cleave

26. Safari sight

27. Kind of first-aid pencil

30. Cousins of crunches

33. Two large muscles of the chest

35. Razor sharpener

37. www.yahoo.com, e.g.

38. Hackneyed

41. "Walking on Thin Ice" singer

42. Ancient Celtic priest

45. Medical exam

48. Overseas

51. Complains

52. Baffled

54. Banquets

55. accident or natural disaster

59. Spoonful, say

62. "God's Little ___"

63. Dostoyevsky novel, with "The"

65. Stronghold captured

66. Synagogue

67. Browning's Ben Ezra, e.g.

68. "I ___ you!"

69. Cozy and comfortable

70. Computer instructions

71. ___ probandi

DOWN

1. Mary in the White House

2. Jewish month

3. Service†club and to promote world peace

4. Maxim

5. A.T.M. need

6. Heroin, slangily

7. Bake, as eggs

8. Ark contents

9. Prevent from entering

10. Not yet final, at law

11. Sundae topper, perhaps

12. European capital

15. Pie cuts, essentially

21. "I'm ___ you!"

23. Aardvark fare

25. Gossip

27. Tater

28. ___ cotta

29. "Wheels"

31. Collection of people or animals or vehicles moving ahead in more†or†less regular formation

32. Navigational aid

34. Back talk

36. Successful runners, for short

39. "___ will be done"

40. Young falcon or hawk

43. With anger

44. ___ any here know me?': King Lear

46. Blue books?

47. Pigment thickly so†that brush

49. Buzzing

50. Fragrant Himalayan

tree

53. Accused's need

55. 100-meter, e.g.

56. Bounce back, in a way

57. Jack-in-the-pulpit, e.g.

58. Arcing shots

60. Balsam used in perfumery

61. Aims

64. ___-tac-toe

PUZZLE 42

ACROSS

1. It is often hampered?

5. Wad in the mouth

9. Big-nosed muppet

14. Word with belly or ear

15. Locale of strike after strike?

16. Eschew

17. It's often polished

18. ___ and terminer

19. Like a ball

20. The 60th wedding†anniversary

23. Origami creature, sometimes

24. 'Curse you, ___ Baron!"

25. Tooth layer

28. Remains to be seen?

30. Recipe instruction

33. Certain variable star

34. Some 1940s internees

35. Free from dampness

36. The 60th wedding†anniversary

40. They're worth three points in Scrabble

41. Certain track auto

42. Meander

43. Word in a supposed Cagney quote

44. Printers' supplies

45. Temporary cessation of something

47. Semicircle, e.g.

48. Bloke's makeshift bed

49. History of the ancient world

56. Mosaic artist's supply

57. Monopoly token

58. Run away

59. Whacked, old-style

60. Safari jacket feature

61. Elbow-wrist connector

62. Was inquisitive

63. Keeps company with

64. Plaster support

DOWN

1. Tinker Bell prop

2. Healthy berry

3. LaBeouf of the 'Transformers' movies

4. One at the wheel

5. Relating to a clone

6. 'The Creation" composer Franz Joseph

7. Yet again

8. Existed

9. Various shrubs and small trees

10. Egg-shaped

11. Common subject?

12. Hurl a barb at

13. Like all prime numbers except 2

21. 'You ___ it to yourself"

22. Not so damp

25. Script direction

26. Bellini opera

27. Salt's ''Halt!"

28. Marital ceremonies, e.g.

29. Kind of friendly?

30. Parting word

31. Motored

32. Metric force units

34. Shaving wound

37. Plucky

38. Skewer

39. Believe or confide readily; full of trust

45. Mortgage closing cost

46. The law, to Mr. Bumble

47. Less than 90 degrees

48. Wild dog of 23-Across

49. Instruments of war

50. Snug corner

51. Pen points

52. Branch headquarters?

53. Southwestern pot

54. Flat payment?

55. One-third of a Fab Four refrain

56. Federal purchasing org.

PUZZLE 43

ACROSS

1. '___ End"

4. Jetted

8. Human or alien

13. Very out-of-pocket

14. Sanctuary

15. Listlessness

16. Causing disapproval or protest.

18. Form of protest

19. No longer in style

20. Hitches successfully

21. 'That's terrible!"

22. Prefix with chemistry

25. Word with spoon or force

26. Joplin composition

28. French pianist and composer (1899-1963)

30. 'Now ___ seen it all!"

31. Room offerer

32. Flammable solvents

34. 'L'___ del Cairo" (Mozart opera)

35. Near in time or place

38. Martin's TV sidekick

39. Fingerboard ridge

40. '... ___ nation under God ...'

41. Pianist's passagework

43. Media solicitations

44. It's after sigma

45. Books that may be cooked

49. Web search result

50. The three to an inning?

51. Tokyo, once

52. Oft-contracted word

53. Jury's determination, perhaps

55. Long-handled spoon

57. "Aida" composer

58. Tiny

62. Anoint, old style

63. Faculty title (Abbr.)

64. Trampled

65. Jewish ceremonial

66. White knight, stereotypically

67. Fractional monetary unit of Japan

DOWN

1. Island united politically with Trinidad

2. Fuji's island

3. God of love

4. Abscam creator

5. Virgo's predecessor

6. Aussie avian

7. Is no longer?

8. To boot

9. 'National Velvet" author Bagnold

10. It's behind closed doors

11. Dennis, to Mr. Wilson

12. Kind of mill

13. Type of book

17. Three-masted vessel

20. Even

23. Not liable to error.

24. Bottom-living cephalopod having a soft oval body with eight long tentacles.

25. Wangle

27. Black fly, e.g.

29. Di-dah lead-in

33. 'You ___ it to yourself"

35. Ark-itect?

36. An indigenous person who was†born in a particular place.

37. Signaled a timeout, say

39. Sailing hazard

41. Artist's studio

42. Adored ones

46. Stand the test of time

47. Ban choice

48. Knight's need

54. Between gigs

56. Is in a cast

57. ___ in victory

58. Letters on a speeding ticket

59. Blood pressure raiser

60. Neither companion

61. Questionable craft

PUZZLE 44

ACROSS

1. Coquettish

4. Summer droners

11. Female demon

16. Amateur video subject, maybe

17. Small particle

18. ...

19. Monarchy occupying most of the Arabian

21. Exploits

22. Combustible heap

23. Nobelist Hammarskjold

24. It's spotted in westerns

25. Colon

31. Parts taken away

34. "C'___ la vie!"

35. #26 of 26

36. Afflict

37. Moray, e.g.

38. At the same time

41. First floor proposal

44. Mozart's "L'___ del Cairo"

45. Hunt for

46. School mos.

48. Decorated, as a cake

52. "The Three Faces of ___"

55. Cleaning floors

58. Chemical reactivity

62. It's a wrap

63. ___ lab

64. Death on the Nile

cause, perhaps

65. Propel, in a way

66. Pertaining to stars

68. Triangular bones

72. 1911 Chemistry Nobelist

73. Cast

74. Arrangement holder

78. Foundation

79. Secretly listen

83. About

84. Tuition classification

85. Archaeological site

86. Beasts of burden

87. Kitchen appliance or best man?

88. Chester White's home

DOWN

1. Astrological transition point

2. White person

3. Part of BYO

4. Modern F/X field

5. Bank offering, for short

6. "Wheels"

7. Wreath for the head

8. Persian Gulf emirate

9. Adjust, in a way

10. Caribbean, e.g.

11. logistics

12. Spikelike inflorescence

13. racial ancestry

14. Chants

15. Balaam's mount

20. Kosher ___

24. Basil-based sauce

26. Gulf of ___, off the coast of Yemen

27. Felt bad about

28. Swallows

29. ___ el Amarna, Egypt

30. "A rat!"

31. Vagabond's wear

32. Cork's country

33. ___ gin fizz

38. Sugar ___

39. "Do ___ others as..."

40. Detective, at times

42. Hawaiian strings

43. Hit the road

47. Balkan capital

49. The "C" in U.P.C.

50. "Empedocles on ___" (Matthew Arnold poem)

51. Indian stewed legume dish (Var.)

53. Viola

54. Circumvent

56. Paul of 'Dinner for Schmucks'

57. Algonquian Indian

58. ___ de deux

59. Depth charges, in slang

60. Wedding acquisitions

61. "___ we having fun yet?"

66. Falls in winter

67. Common request

69. Liquid excretory product

70. Safari sight

71. Anatomical sac

75. Boosts

76. Agitated state

77. Like custard

78. Bleat

79. 1999 Pulitzer Prize-winning play

80. Sylvester, to Tweety

81. Absorbed, as a cost

82. The "p" in m.p.g.

PUZZLE 45

ACROSS

1. Endure

5. Caper

11. Accused's need

16. "How ___ Mehta Got Kissed, Got Wild, and Got a Life" (Kaavya Viswanathan novel in the news)

17. Park, for one

18. Physics lab device, for short

19. Ore or coal

21. Large lemur

22. Towers over the field

23. "Beetle Bailey" dog

24. Time in power

25. Hate, say

27. Urine

29. Toni Morrison's "___ Baby"

30. A mistake

33. Chocolate and cream

36. #26 of 26

37. Knight stalker

39. "___ bitten, twice shy"

41. Fungal spore sacs

45. West Indies

49. Freshman, probably

50. Assortment

51. Assert without proof

52. Clobber

54. Part of a board

56. Represent falsely

61. When it's broken, that's good

64. Native or inhabitant

65. Call to a foxhound

68. Beachwear

69. Vex, with "at"

72. "Hurray!"

73. City on the Arkansas River

74. A measuring instrument

76. Arab leader

77. Indian turnover

78. "And ___ thou slain the Jabberwock?"

79. Tears

80. Catkins

81. Bad day for Caesar

DOWN

1. Baby

2. Ideals

3. Charades, e.g.

4. "Four Quartets" poet

5. Video maker, for short

6. Of or relating to avionics

7. Blow off steam

8. "Don't bet ___!"

9. Expire

10. Two-year-old sheep

11. Property recipient, at law

12. Hindu phallic

13. Produce a literary work

14. They mix and serve

15. Worst for driving

20. Telekinesis, e.g.

24. Baptism, for one

26. Certain Arab

28. Number of decks

31. Bar

32. Flat braided cordage

34. Not "fer"

35. "Smoking or ___?"

37. "Silent Spring" subject

38. ___ v. Wade

40. 7 zeros; ten million

41. ___-bodied

42. A sleepy person

43. Butt

44. "Rocks"

46. Dried coconut meat

47. Ziti, e.g.

48. Final: Abbr.

52. Begging

53. Sonata, e.g.

55. Loosen, as a cap

56. Concern

57. Bury

58. Hot

59. Finished cleaning

60. Memory trace

62. Quip, part 2

63. Places to sleep

66. Alkaline liquid

67. Police in India

70. Finger, in a way

71. Shakespeare, the Bard of ___

74. Fed. construction overseer

75. More, in Madrid

PUZZLE 46

ACROSS

1. Cabinet acronym, once

4. Concrete type

11. Batter's position

16. Absorbed, as a cost

17. Otalgia

18. Kidney enzyme

19. Unsaponified fat

21. ___ out (declined)

22. ___ bread

23. "___ any drop to drink": Coleridge

24. Daisylike bloom

25. Electromotive force

31. Suppress, in a way

34. Calypso offshoot

35. Born, in bios

36. Bauxite, e.g.

37. Cal. col.

38. Wheedle

41. From now on

44. Always, in verse

45. Santa ___, Calif.

46. Water wheel

48. Component used in making plastics and fertilizer

52. "Arabian Nights" menace

55. Jaded

58. Agree on a contract

62. ___ few rounds

63. Fed. construction overseer

64. "The ___ Daba Honeymoon"

65. Bit

66. Some canines

68. Systems analysis

72. Beats it

73. Beaver's work

74. Beer buy

78. Spikelike inflorescence

79. Party bowlful

83. "Tootsie" Oscar winner

84. Mint family member

85. Pickpocket, in slang

86. ...

87. Make infertile

88. Chester White's home

DOWN

1. Door fastener

2. Decorative case

3. Cried

4. The "p" in m.p.g.

5. Churchill's "so few": Abbr.

6. Victorian, for one

7. Persian attraction

8. Follow, as a tip

9. Dig discovery: Var.

10. "First Blood" director

Kotcheff

11. Gland in men

12. Greek penny

13. Feeler

14. Male hawk

15. Armageddon

20. "___ of Eden"

24. tropical fruit

26. "... or ___!"

27. Book part

28. Sleep on it

29. "It's no ___!"

30. "A rat!"

31. 1922 Physics Nobelist

32. Sundae topper, perhaps

33. Addition column

38. Razor sharpener

39. Blender sound

40. "How ___!"

42. Diamond, e.g.

43. Column crossers

47. Star in Perseus

49. Anger

50. At one time, at one time

51. Indian maid

53. Performers or singers

54. Bucks

56. Motherless

57. Telephone line acronym

58. Consumes

59. Really bad

60. Emerging

61. Infomercials, e.g.

66. Kind of control

67. Carve in stone

69. Argentine dance

70. Dig, so to speak

71. Singers Ruess and Dogg

75. Boosts

76. Fast-moving card game

77. Catch a glimpse of

78. ___ grecque (cooked in olive oil, lemon juice, wine, and herbs, and served cold)

79. Detachable container

80. Burden

81. "___ moment"

82. Atlantic catch

PUZZLE 47

ACROSS

1. Ado

5. Computer dialers

11. Accumulate

16. ...

17. Footless

18. Fancy home

19. A preliminary discussion

21. First name in mystery

22. Good substantial quality

23. Peewee

25. Opening time, maybe

26. Bidder receives points toward game

29. 50 Cent piece

32. "For shame!"

33. Be a rat

34. Slips

37. Two-year-old sheep

40. "The English Patient" setting

44. Like twilight; dim

47. Bring (out)

48. Neat; spruce

49. Dirt

50. English race place

51. Appraiser

52. Give individual character to

54. Instruction to go away

57. Clinton, e.g.: Abbr.

58. 1,000-kilogram weights

59. Cheat

61. Fed. construction overseer

63. "A jealous mistress": Emerson

64. A sale to reduce inventory

71. H.S. subject

72. Blue hue

73. Ill temper

77. Old Roman port

79. Transparent and opaque objects

81. Bread spreads

82. It may save your skin

83. Not kosher

84. Joanna of 'Growing Pains'

85. Bad looks

86. Heated phone message?

DOWN

1. Easy dupes

2. Novice

3. "American ___"

4. Archaeological find

5. Purplish red colors

6. City 175 miles north of Lisbon

7. ...

8. Swelling

9. Bouncing off the walls

10. Biases

11. "___ Maria"

12. ...

13. Cold

14. What "yo mama" is

15. Eastern wrap

20. Tokyo, formerly

24. ...

27. Election data

28. Biochemistry abbr.

29. Reconnaissance

30. Ravine

31. Future M.D.'s course

35. Official approval

36. "For shame!"

38. Say "Li'l Abner," say

39. The act of playing for stakes

41. Secrets

42. Fan

43. Bear witness

45. Modern F/X field

46. Gun, as an engine

50. "Much ___ About Nothing"

53. Neighbors of the Sammarinese

55. Alpine sight

56. Apartment

60. Gets down

62. Knight in shining armor

64. ...

65. Fine thread

66. Banana oil, e.g.

67. Poultry buy

68. A-list

69. "Star Trek" rank: Abbr.

70. Divisions

74. Blood's partner

75. Crown

76. Steal goods; take as spoils

78. Balaam's mount

80. Conk out

PUZZLE 48

ACROSS

1. The first stage of meiosis

9. Battery contents

13. Discouraging words

16. Signal amplifier

17. Centers of activity

18. Egg cells

19. Physical and biological aspects of the seas

21. After expenses

22. "___ Gang"

23. Fancy-schmancy

24. Appear

25. Grand ___ ("Evangeline" setting)

26. Swiss capital

29. The act of concealing yourself

32. Song and dance, e.g.

35. Chest protector

36. "___ Brockovich"

37. Source of energy

42. Boys

43. Altar avowal

44. Wouldya like to?'

45. ___ el Amarna, Egypt

46. A republic in the West Indies

53. Balaam's mount

54. Deceive

55. Victorian, for one

56. "Empedocles on ___" (Matthew Arnold poem)

58. The father of your spouse

62. Intro to physics?

63. ___-Wan Kenobi

64. Bang-up

65. Having incalculable monetary

68. Dinner signals

72. Mythical monster

73. Cambridge sch.

74. Kind of column

78. 1969 Peace Prize grp.

79. Chinese dynasty

80. Maximum speed and firepower

84. "Star Trek" rank: Abbr.

85. Biology lab supply

86. Goods carried

87. "Losing My Religion" rock group

88. Bank

89. Forward spin

DOWN

1. 86 is a high one

2. Come again

3. "La BohËme," e.g.

4. ___ green

5. ___ Solo of "Star Wars"

6. Above

7. ___ lily

8. Blows it

9. Matterhorn, e.g.

10. Live together

11. Bar 'rock'

12. Potting soils, e.g.

13. The sequential performance of multiple operations

14. Went too far with

15. Glossy fabrics

20. "Bingo!"

27. Apprehend

28. Derby parts

30. "Dilbert" cartoonist Scott Adams has one: Abbr.

31. People person

33. Asian tongue

34. Harmony

37. Massenet's "Le ___"

38. "Much ___ About Nothing"

39. Romantic ideals and attitudes

40. ___ lab

41. Russian alternative

45. Calibrating something

47. Bartender on TV's Pacific Princess

48. ___ v. Wade

49. "The Snowy Day" author ___ Jack Keats

50. 100 centavos

51. Bank offering, for short

52. Aviary sound

56. Authorize

57. Earthly

58. Leaves

59. The verbal act of urging on

60. "___ the season ..."

61. "The Matrix" hero

66. Hinder

67. Costa del ___

69. Japanese-American

70. Occupant of Friendship 7

71. Alphabetizes, e.g.

75. Fly, e.g.

76. Bounce back, in a way

77. Big bore

81. "Don't give up!"

82. Bull markets

83. Babysitter's handful

PUZZLE 49

1	2	3	4		5	6	7	8	9		10	11	12	13	14	15
16					17						18					
19			20								21					
22					23					24						
25				26			27	28								
29				30			31					32	33	34	
35				36			37				38				
		39			40					41				
	42	43		44				45	46				
47				48				49					
50			51	52			53				54	55	56
57			58				59			60			
		61				62			63			
64	65	66				67			68			
69					70			71				
72					73					74			
75					76					77			

ACROSS

1. Boutique

5. Big dipper

10. Boards

16. "I, Claudius" role

17. Arab leader

18. Cinch

19. In an Italian style

21. Spooks

22. Place for a barbecue

23. .0000001 joule

24. Minimum

25. Explodes when struck

29. Ashtabula's lake

30. Absorbed, as a cost

31. Advocate

32. Marienbad, for one

35. Burden

36. Chester White's home

37. Feeling nausea

39. Blue hue

40. ___ souci

41. Court attention-getter

42. Covered wagon

47. Drudgery

48. Neural network

49. Deception

50. A state of misfortune or affliction

53. "Cool" amount

54. Software program, briefly

57. "It's no ___!"

58. Household chore

59. Kills the helper T cells

60. Algonquian Indian

61. Circular firework

64. A complex inorganic compound

67. Charlotte-to-Raleigh dir.

68. Garlicky mayonnaise

69. Haitian monetary unit

70. Makes wet and dirty, as from rain

72. "The Wizard of Oz" prop

73. Of the vascular layer of the eye

74. Boat propellers

75. Extracts

76. Actress Oberon

77. European language

DOWN

1. Scrap

2. A female paramour

3. A woman plaintiff

4. Watch over

5. Grassland

6. Forgiveness of a sort

7. Honeybunch

8. "Take your hands off me!"

9. "... ___ he drove out of sight"

10. Blowhards

11. Coop flier

12. Autocrats

13. "No problem!"

14. Assayers' stuff

15. Home, informally

20. Chit

26. Debaucher

27. Shade

28. Infatuation

32. Eye affliction

33. Equal

34. Cutting tool

36. Schuss, e.g.

37. Foul

38. Bang-up

39. Ad headline

40. Boil

42. Miniature sci-fi vehicles

43. Bank of Paris

44. Dander

45. Antipasto morsel

46. Black gold

47. ___ cross

51. Kigali resident

52. Glossy fabrics

53. 20 Questions category

54. Like an iris part

55. Kitchen gadgets

56. A sleeveless cape with fur

59. Impede

60. Small tropical flea

61. About

62. "The Canterbury Tales" pilgrim

63. Card

64. All excited

65. Churn

66. Stubborn beast

70. Depress, with "out"

71. A pint, maybe

PUZZLE 50

ACROSS

1. Italian ice cream

8. Strong liquors

15. Hunks

16. Approach with stealth

17. Separated

18. Apply again

19. Lover with a ladder, perhaps

20. John and others

21. M√°laga man

22. Hires competition

23. Blue hue

25. Twangy, as a voice

27. Threadbare

28. Beyond calculation or measure

33. "Rocky ___"

34. Barely beat

35. Resolve

36. "Smoking or ___?"

37. In-flight info, for short

38. Expose to cool or cold air

40. Victorian, maybe

42. Accustom

43. Western Samoa money

44. [Just like that!]

46. Freshwater fishes

50. Giving the sensation of tension

52. Outcast

53. Corrupt morally or by intemperance or sensuality

54. Rap variety

55. Figure

56. Bromo ingredient

57. Eternal

58. Class work

DOWN

1. Checks out

2. Big name in computers

3. Sage

4. Comfy shoes

5. To cure or restore

6. ___ de force

7. Death on the Nile cause, perhaps

8. New England catch

9. Salad green

10. Bucket of bolts

11. Catmint

12. A cocktail of vodka

13. A sharp transient wave in the normal electrical state

14. 007, for one

20. White meat mold

22. The Talented Mr. Ripley' star

24. Fabricator's forte

26. "Your majesty"

27. Any agent that retards

28. All thumbs

29. Treat with nitric acid

30. Shoulder board (Var.)

31. Spring sound

32. Corker

39. Mob disperser

41. 1973 Elton John hit

44. Calamine targets

45. Certifies

47. Kid's name

48. Dine at home

49. Some food fishes

51. Dressing ingredient

52. French door part

53. ___ Appia

54. "For Me and My ___"

PUZZLE 51

ACROSS

1. Asian nurse

5. Telekinesis, e.g.

8. Computer architecture acronym

12. Opera star

13. Five spots on a five-spot

14. Andrea Doria's domain

15. Soon, to a bard

16. Related to the anus

17. Synthetic resin

18. Spent only for a particular purpose

20. "Naked Maja" painter

21. Put to rest, as fears

22. Calendar abbr.

23. Stellar

26. Repudiate

30. Dash abbr.

31. Wicker material

34. Container weight

35. ...

37. ___ grass

38. Arctic

39. Brass component

40. Gooey cake

42. Chill (out)

43. Brainiac

45. Cause of hereditary variation

47. "Spy vs. Spy" magazine

48. Symmetric crystal

50. Cicatrix

52. Hoofed mammals having very thick skin

56. Recipe direction

57. A chip, maybe

58. Greek earth goddess: Var.

59. Bad for dieters

60. Animal with a mane

61. Catch a glimpse of

62. At the home of

63. Ring bearer, maybe

64. Bakery selections

DOWN

1. Jewish month

2. Peewee

3. Shakespeare, the Bard of ___

4. Phantom home?

5. Feather, zoologically

6. Out there

7. "Cast Away" setting

8. Set up new headquarters

9. Black

10. ___ bean

11. Blackguard

13. Spanish dish

14. Light green plums

19. Danger signal

22. Kipling's "Gunga ___"

23. Astound

24. Bit of parsley

25. Flip-flop

26. Anniversary, e.g.

27. Brass button?

28. Architectural projection

29. Piece cut from a cheese wheel

32. Contemptible one

33. Sylvester, to Tweety

36. Sentimentality in art or music

38. unning water

40. "Crikey!"

41. German cathedral city

44. Before the due date

46. Record holder?

48. Craze

49. Gibson, e.g.

50. Literally, "king"

51. Commend

52. Blanched

53. "Duck soup!"

54. Opportune

55. 1951 N.L. Rookie of the Year

56. A fluorocarbon with chlorine

PUZZLE 52

ACROSS

1. Smock

6. Woolly bear, eventually

10. Monk's hood

14. The English translation for the french word: apnÉe

15. Chad's place

16. Teatro ____ Scala

17. Cleanup of a sort

20. Theologian's subj.

21. Silver ____

22. Leader of the Medicine Show, in 1970's rock

23. WB competitor

25. Voice of America org.

26. Things Aristotle wrote

33. Wit's end?

34. Covered again, as an air route

35. Fill in the blank with this word: "____ regni (in the year of the reign)"

36. Standing by

38. Fill in the blank with this word: "Debussy's "Air de ____""

39. Russia's Lake ___

40. Teetotalers

41. Fill in the blank with this word: "___ Volcanic National Park"

43. See 20-Across

44. Make attractive requests?

47. Fill in the blank with this word: ""All ___ are off""

48. Popeye's Olive ___

49. Some Girl Scout cookies

52. Popular brew, for short

54. Fill in the blank with this word: ""Should ___ shouldn't ...""

57. Boxing legend

61. Fill in the blank with this word: "___ Rabbit"

62. New Rochelle college

63. Fill in the blank with this word: "___ and aahed"

64. Unscramble this word: rayd

65. Nevada's state tree

66. Victorian, maybe

DOWN

1. Youth

2. Takeoff artist

3. Fill in the blank with this word: "___ in a blue moon"

4. Wind god

5. Fill in the blank with this word: "Farmer's ___"

6. Picasso's muse Dora ___

7. You'll be the death ___!'

8. Cable TV giant

9. Woodcutter's tool

10. Native of old China

11. Fill in the blank with this word: "Aglio e ___ (pasta dressing)"

12. Defeat

13. Thin

18. Unit for a lorry

19. Seed covering

24. Whittle down

25. Restless

26. Never, in Nogales

27. Togetherness

28. They're found by the C's

29. Fill in the blank with this word: ""He's making ___ and checking ...""

30. 1983 Indy 500 winner Tom

31. "Picnic" playwright's kin

32. Zoom

33. Some camp denizens, for short

37. China's place

39. Right turn ___

41. Speaks without thinking, perhaps

42. Lyric poem

45. Word before window or end

46. Romance novelist ___ Glyn

49. Is loyal to

50. Fill in the blank with this word: ""___ Lee" (classic song)"

51. Hate or fear follower

52. Fill in the blank with this word: "___ means (not at all)"

53. Fill in the blank with this word: "___-Ude (Trans-Siberian Railroad city)"

55. Where Pearl City is

56. Snow White's sister

58. Get an ___ (ace)

59. Fill in the blank with this word: "Alley ___"

60. Potus #34

PUZZLE 53

ACROSS

1. I...

4. All excited

8. Boat with an open hold

12. Bunk

15. "Rocks"

16. Hindu Mr.

17. Voltage through motion

19. Chinese restaurant offering

21. Logician

22. Begin

23. Extreme oldness

25. "___ Gang"

26. Coal carrier

27. Pandowdy, e.g.

28. Absorbed, as a cost

29. Productive period

32. 2004 nominee

34. Celebrations

36. Informal term

39. Incorrect in behavior

40. Branch

41. Climbing bean or pea plant

42. Antiquity, in antiquity

43. Certain protest

45. Blubber

46. "ER" network

48. Boat in "Jaws"

52. 1999 Pulitzer Prize-winning play

54. See-through sheet

57. Long, long time

58. Bud holders?

61. Aggressive

63. Animal hides

64. "Go on ..."

65. Field

66. "___ we having fun yet?"

67. Amigo

69. ___ Today

72. Modern F/X field

73. straight line or lines

76. 1,000 kilograms

79. Power and authority

80. Commits sabotage

82. Security checkpoint

83. Antares, for one

84. Schuss, e.g.

85. Oolong, for one

86. Change

87. Cravings

88. "Comprende?"

DOWN

1. Butts

2. "God's Little ___"

3. ...

4. Eccentric

5. Gangster's gun

6. Witchcraft and sorcery

7. Sea birds; used as fertilizer

8. ...

9. Annoying and unpleasant

10. Certain Arab

11. Inefficient in use of time and effort and materials

12. Narrowly triangular, wider at the apex

13. Clytemnestra's slayer

14. Minimally worded

18. Elephant's weight, maybe

20. Ladies' bag

24. Kind of drive

29. Org. that uses the slogan 'Aim High'

30. Legal prefix

31. Gets rid of

33. Administrative unit of government

35. Edible starchy

37. "I" problem

38. Irish Republic

39. The terminal section of the alimentary canal

44. Blockhead

47. High school class, for short

49. Inclination

50. Chanel of fashion

51. Again

53. Who stimulates and excites people

55. Excessive

56. A muscle which raises any part

58. Most economical

59. Phormio' playwright

60. Euripides drama

62. A homosexual man

64. Parallel or straight

68. "Home ___"

70. Flip, in a way

71. Winged

74. Not just "a"

75. "I, Claudius" role

77. Microwave, slangily

78. Ashtabula's lake

81. Blackout

PUZZLE 54

ACROSS

1. Crack, in a way

5. Refuse

10. 100 kurus

14. Maneuvered a ship

15. Accept

16. Carbon compound

17. "O" in old radio lingo

18. Disregard

20. Oater fellow

22. Oolong, for one

23. "It's no ___!"

24. Breed

25. Women only

27. Five Nations tribe

31. Animal house

32. Literally, "dwarf dog"

33. Misfortunes

35. Rear

39. Banded stone

40. Shaggy Scandinavian rug

41. Anatomical dividers

42. Aria, e.g.

43. Brio

44. Santa's reindeer, e.g.

45. Nth degree

47. People are cared for

49. Desk accessory

53. "Bingo!"

54. Control

55. Propel, in a way

56. Spray very finely

60. Socially incorrect

63. Dugout, for one

64. Accommodate

65. Contemptuous look

66. "Empedocles on ___" (Matthew Arnold poem)

67. Coastal raptor

68. Overhangs

69. Beams

DOWN

1. Bonbon, to a Brit

2. Bindle bearer

3. Acknowledge

4. Enjoys seeing sex

5. Polo match

6. Better

7. Mandela's org.

8. Specialty

9. Hit sitcom

10. "Fantasy Island" prop

11. Bring upon oneself

12. Certain tribute

13. Back street

19. Bind

21. Any Platters platter

25. Raise sail or flag

26. Waiting area

27. Wood sorrels

28. Canceled

29. Face-to-face exam

30. Demoiselle

34. Channel

36. "What've you been ___?"

37. Check

38. "Unimaginable as ___ in Heav'n": Milton

41. Group of Bantu languages

43. Stretch

46. Long, long time

48. Don Juans

49. Cursor mover

50. Dog tag datum

51. Deprive of courage

52. "Gladiator" setting

56. On the safe side, at sea

57. Bit

58. Buffoon

59. Flight data, briefly

61. Absorbed, as a cost

62. Gun, as an engine

PUZZLE 55

ACROSS

1. Branch

4. An aromatic exudate from the mastic tree

11. Fergie, formally

16. Morgue, for one

17. Snakes

18. Chip away at

19. Habitual

21. Crosses

22. Bridges of Los Angeles County

23. Elmer, to Bugs

24. Amiens is its capital

25. The chief mountain range of western North America

31. An orange isomer

34. "Come to think of it ..."

35. Tokyo, formerly

36. Trick taker, often

37. 20-20, e.g.

38. Give birth

41. A broad cartridge belt

44. Affranchise

45. Locale

46. Cake part

48. "The Last of the Mohicans" girl

52. Head, for short

55. A closed litter

58. First to think of or make

62. Conk out

63. Engine speed, for short

64. Popular fish for show

65. Blue

66. Tusked mammals

68. Climbing over, crawling through,

72. Cupid's boss

73. Bull markets

74. "___ of Eden"

78. Burgundy grape

79. A curved section or tier

83. Anoint, old style

84. Eat at a restaurant or at somebody else's home

85. "Spy vs. Spy" magazine

86. Stage item

87. Hard-to-call contests

88. Dash lengths

DOWN

1. Kuwaiti, e.g.

2. Bumpkin

3. ___ Verde National Park

4. Back-to-work time: Abbr.

5. Appropriate

6. "___ Cried" (1962 hit)

7. 60s coloring method

8. Going to the dogs, e.g.

9. Kid's name

10. "Casablanca" pianist

11. Microorganisms or viruses

12. Bouquet

13. Having ample space

14. Extras

15. "For ___ a jolly ..."

20. 100 cents

24. Tangle

26. "Beetle Bailey" dog

27. Provide with an overhead surface

28. Show respect, in a way

29. "It's no ___!"

30. Costa del ___

31. Street fleet

32. Healthy berry

33. Cost of living?

38. One with a crowbar, perhaps (Var.)

39. Brought into play

40. Carve in stone

42. Kosher ___

43. "___ on Down the Road"

47. "M*A*S*H" role

49. Boat propellers

50. Opportune

51. Kind of dealer

53. Excited in anticipation

54. Kind of concerto

56. Lagerlˆf's "The Wonderful Adventures of ___"

57. Part of a bird's beak

58. Alias

59. "Semiramide" composer

60. Infusion of e.g.

61. 100 lbs.

66. Weak and ineffectual

67. ___-friendly

69. American chameleon

70. Antique shop item

71. Some tournaments

75. #1 spot

76. Brickbat

77. "Bill & ___ Excellent Adventure"

78. When it's broken, that's good

79. "Silent Spring" subject

80. Former French coin

81. Trophy

82. "___ alive!"

PUZZLE 56

1	2	3	4		5	6	7	8	9		10	11	12	13
14					15						16			
17					18						19			
20			21						22					
23						24					25	26	27	
		28		29	30	31			32					
33	34	35		36			37	38						
39			40		41						42			
43			44					45		46				
47					48				49					
50				51	52			53		54	55	56		
		57			58	59	60							
61	62	63		64					65					
66				67					68					
69				70					71					

ACROSS

1. Dead-end jobs, e.g.

5. Bars or bolts

10. Shark variety

14. Adjoin

15. Danger

16. Decorated a cake

17. Speck of dust

18. Marine biology subject

19. Eye annoyance

20. Defender of some group or nation

23. Police trap

24. Versus

28. Drinks alcohol to excess habitually

32. Grassy plain of Latin America

33. Baseball great Mel

36. To get in or get out

39. Mercury and Saturn, but not Earth

41. Like poltergeists

42. There's probably money in it

43. 365 (or 366) days

46. Bloom-to-be

47. Very, to the maestro

48. Artifices

50. Ones with iron hands

53. Hitches successfully

57. Nonsense

61. Aerated beverage

64. Norse love goddess

65. German leader Helmut

66. 'A Prayer for ____ Meany"

67. Let go

68. Fanzine focus

69. Push a product

70. Feet are in them

71. McGee's closet, e.g.

DOWN

1. Boat launches

2. WWII sub

3. 'Frutti" intro

4. Some skeletal parts

5. Reach across

6. Tailor-made items

7. Fertilizer ingredient

8. Knee-to-ankle bone

9. It's the word on the street

10. Forcibly thrown or projected

11. Get off the fence

12. Anthem author

13. Shelley praise

21. Check out shamelessly

22. Powder substance

25. Mogul empire

26. Major mix-up

27. Trifled (with)

29. It might be skinned in the fall

30. Checklist unit

31. One way to get to first base

33. Eightsome

34. Leather strap

35. Portion of hair

37. Indigenous Japanese people

38. Clears, on a pay stub

40. Directly or immediately

44. Old orchestral string

45. For the missus

49. Former kingdom in northeastern India

51. It may be pulled at a carnival

52. Narrow channel

54. Semiconductor, perhaps

55. Distinctive spirit of a culture

56. Flies off the shelf

58. Title for Helmut Kohl

59. Word with green or eagle

60. X-ray units

61. One with a beat?

62. Symbol of wisdom

63. Money of Romania

PUZZLE 57

ACROSS

1. Obtained from urine

5. Double-reed woodwind

9. Elongated fruit (Var.)

14. Sacred Hindu writing

15. Prison sentences

16. Cover story?

17. Make uniform

18. Intl. commerce pact

19. Skirt fold

20. Australia's national day

23. Distinctive spirit of a culture

24. Relating to the ear

25. 'That's all ___ wrote"

28. Protects the body from foreign substances

31. Pint-size

34. Is not well

35. Of the first water

36. Supplement

38. Not likely to get by the censor

41. Scurries

42. Civil aviation

43. Golf standard

44. From time to time

49. Baseball great Mel

50. Current jumps, e.g.

51. Loamy deposit

54. Friend or competitor

57. Grape seeds

60. Invited a perjury charge

61. Lion's pride

62. Rainwater pipe

63. Strongly suggest

64. Aquarium dweller

65. Flower from the violet family

66. Famous septet

67. Where worms may be served

DOWN

1. Part of the eye

2. Las Vegas show, perhaps

3. Cato's clarification

4. Upper and lower eyelids meet

5. Very pleasurable

6. False god

7. Alternatives

8. Beverly Hills home, typically

9. Church rule

10. A good friend indeed

11. Dessert favorite

12. Arab overgarment

13. Funny one

21. Like some numerals

22. Insult, slangily

25. Set of steps

26. Redhead's secret

27. Islamic bigwig

29. Farthest or highest (Abbr.)

30. Marsupial pocket, e.g.

31. "Yippee!"

32. ___ of Nantes, 1598

33. DVD player button

37. NATO member

38. 'L'___ del Cairo" (Mozart opera)

39. 3 stanzas and an envoy

40. Wise legislator

42. Fire up

45. Marbles you don't play with

46. Blood pressure raiser

47. The organ of sight

48. The boss' "echo"

52. Stratified rock

53. Joins the choir

54. Bothersome burden

55. Hit prefix

56. Bog fuel

57. Egyptian cobra

58. Bookkeeper

59. Charged particle

PUZZLE 58

1	2	3	4	5		6	7	8	9		10	11	12	13
14						15					16			
17						18					19			
20					21						22			
				23					24	25				
26	27	28	29		30				31					
32					33			34		35		36	37	38
39				40				41	42					
43						44	45				46			
		47		48		49				50				
51	52	53				54			55					
56					57	58					59	60	61	62
63					64					65				
66					67					68				
69					70					71				

ACROSS

1. Philippine banana plant

6. Famous brother

10. European freshwater fish

14. Trivial Pursuit edition

15. Wolfe, the sleuth

16. Penultimate word in a fairy tale

17. 'Bellefleur" author

18. Cream was one

19. Expurgate, editorially

20. A detailed plan

22. Vena --- (vessel to the heart)

23. Native American tent

24. Nonpointed end

26. Hit alternative

30. Networked computers, for short

31. Pencil stump

32. Agency concerned with civil aviation

33. Kim Jong-il's place

35. Type of iron girder

39. Undone or leave†out

41. Pay for, as a project

43. Father of Jacob

44. Something fishy?

46. Word with "movie" or "party"

47. Pythagorean P

49. Romantic or Victorian, e.g.

50. Where starter

51. Berry the size of a hen egg

54. Kind of blocker

56. Endangered buffalo

57. China or silverware

63. Common subject?

64. Dangerous marine creature

65. Subject of a house inspection test

66. Between gigs

67. Fruit with a wrinkled rind

68. Not under one's breath

69. Holding place

70. Hammered at a slant

71. Puppies' cries

DOWN

1. Slack-jawed

2. Gal's sweetheart

3. Contrary one

4. Handed a line

5. Balance sheet plus

6. Italian meal starter, perhaps

7. Italian sculptor

8. Albany canal

9. Undo, in a way

10. Head of purplish-red leaves

11. Eye layers

12. Dig deeply

13. Bleak, in verse

21. Lavender flower

25. Remains to be seen?

26. Abbreviated version

27. Tries to reduce swelling, in a way

28. Long story

29. Electrical device

34. Raised above?

36. Organic compound

37. Blackjack components

38. Word with pittance

40. Canyon sound

42. Hot under the collar

45. Humiliating failure

48. Decline an invitation

51. Frenzied

52. Battery terminal

53. '___ be sorry!"

55. Arrangement

58. Thus

59. Low tract

60. Fanzine focus

61. ___ d'etat

62. Wraps up

PUZZLE 59

ACROSS

1. Bohemian, e.g.

5. Accomplishment

9. Permeate

14. ___ of the above

15. Annul

16. ...

17. From a foreign country

19. Lid or lip application

20. See if or how it works

21. National Zoo favorites

22. Nod, maybe

23. Reprimand, with "out"

24. Yellowish pink

28. Butt

29. "Awright!"

33. Rainbow ___

34. Exude

35. Ballad

36. Normal temperature of room

40. Person in a mask

41. Medical advice, often

42. Deceived

43. Poet Angelou

45. "Act your ___!"

46. Lead source

47. ___ Verde National Park

49. Keep out

50. Advantages

53. Soup cooked in a large pot

58. Balloon probe

59. A windstorm

60. Geometrical solid

61. On the safe side, at sea

62. History Muse

63. Anatomical dividers

64. Frau's partner

65. Suspended

DOWN

1. Arrogant and annoying

2. Look angry or sullen

3. The "A" of ABM

4. A constellation in the southern hemisphere

5. Charity event in the park

6. Provide, as with a quality

7. Gulf of ___, off the coast of Yemen

8. Anderson's "High ___"

9. Acquired relative

10. For the most part

11. Fasten

12. Component used in making plastics and fertilizer

13. All ___

18. "Jo's Boys" author

21. Cell alternative

23. "Carmen" composer

24. Play, in a way

25. Bouquet

26. Eccentric

27. Ornamental flower, for short

28. Small woods

30. Avoid

31. Composer Copland

32. Howler

34. Alpha's opposite

37. Clear, as a disk

38. Dismays

39. ___ el Amarna, Egypt

44. During

46. Elastic

48. Swelling

49. More despicable

50. Express Mail org.

51. Dermatologist's concern

52. Barber's motion

53. Stubborn beast

54. Allergic reaction

55. Yellowish brown balsam

56. "___ Brockovich"

57. E.P.A. concern

59. Code word

PUZZLE 60

1	2	3	4			5	6	7	8	9		10	11	12
13				14		15						16		
17					18							19		
20							21			22				
			23			24	25		26					
27	28	29		30			31		32					
33				34				35		36		37	38	
39			40				41		42					
43				44		45					46			
		47		48		49					50			
51	52			53		54			55					
56					57			58			59	60	61	
62				63			64	65						
66				67					68					
69				70						71				

ACROSS

1. Medieval chest

5. Fill in the blank with this word: "___ Nurmi, the Flying Finn"

10. Finish this popular saying: "Time and tide wait for no_____."

13. Shell lining

15. Imprison

16. Formula ___

17. Phone calls, room service charges, etc.

19. Wind dir.

20. The English translation for the french word: Malawi

21. Jones and Smith, maybe

23. slang for sexual intercourse

26. Synchronized (with)

27. Fill in the blank with this word: "'"___ bad!'"

30. Sale table notation

32. The Everlys' "When Will ___ Loved"

33. Scottish Peace Nobelist John Boyd ___

34. Characters in 'Romola' and 'The Gondoliers'

36. Vexes

39. Garfield's assassin

41. 1982 Grammy-winning singer for "Gershwin Live!"

43. Have ___ in one's bonnet

44. The Braves' div.

46. 37-Across's age on May 29, 2003

47. Wichita-to-Omaha dir.

49. Worse than jitters

50. Fill in the blank with this word: ""Star Trek: ___""

51. What the majority of elements are

54. Fill in the blank with this word: ""I call 'em as I ___""

56. Nudist

58. Violent, perhaps

62. What's right in front of U

63. Something a company won't reveal

66. Star of "Youngblood," 1986

67. The English translation for the french word: chevalet

68. Nothing runs like a ___' (ad slogan)

69. Switch ups?

70. Show with Jean-Luc Picard as captain of the Enterprise, in fan shorthand

71. Fill in the blank with this word: ""So ___?""

DOWN

1. Spiritedly, in music: Abbr.

2. Fill in the blank with this word: "___ temperature (was feverish)"

3. Multiple of XXXV

4. Met highlights

5. Writer

6. Quantity: Abbr.

7. Simple rhyme scheme

8. Singer with a falsetto

9. Noble family of medieval Italy

10. Halloween success story?

11. Fill in the blank with this word: ""Bonne ___!" (French cry on January 1)"

12. Items sold at stands

14. First American to walk in space

18. Resident of Asmara

22. The same size

24. Fill in the blank with this word: "Author ___ Le Guin"

25. Zoologist's foot

27. The English translation for the french word: toge

28. Day spa offering

29. Gives freshman introduction, say

31. Forced feeding, as with a tube

35. Impertinent types

37. Particle named for a letter of the alphabet

38. Make out, to Harry Potter

40. Time for a coffee break, maybe

42. Pronounced

45. U.S.N.A. grad

48. In groups

51. Thomas of "That Girl"

52. PelÈ's real first name

53. Fill in the blank with this word: "___-O-Matic (maker of sports games)"

55. Uses an airborne defense

57. White House's ___ Room

59. Smartphone introduced in 2002

60. Fill in the blank with this word: "___ Aarnio, innovative furniture designer"

61. Call in the game Battleship

64. Opium ___

65. Taina who was one of Les Girls, 1957

PUZZLE 61

ACROSS

1. Sue Grafton's '___ for Evidence'

4. The English translation for the french word: acide ?-linolÈnique

7. Quicken Loans Arena cagers

11. You never had ___ good!'

12. ___ deck

13. Scottish land-owner

15. "Wow, congrats!"

17. Knife brand

18. Vardalos of the screen

19. They worked in Sparta

21. Tin ___

22. Line of work: Abbr.

23. Wing: Abbr.

24. Year of the last known Roman gladiator competition

27. Presidential inits.

28. Mark your card!

30. Here ___, there...' ('Old

MacDonald' lyric)

33. Waters, informally

36. Tahitian-style wraparound skirt

38. Red indication on a clock radio

39. Indian poet ___ Aurobindo

40. Compass points (seen spelled out in 20-, 26-, 43- and 53-Across)

41. They're entered in court

43. Band with the 1988 #1 hit "Need You Tonight"

45. Fill in the blank with this word: ""___ show you!""

46. They're below some chests

48. Year in Septimius Severus's reign

50. Temperate

51. Silver coin of ancient Greece

53. Reno and Kennedy, e.g.: Abbr.

56. Org. for Venus and Serena Williams

58. Italian astronomer Giovanni Battista ___, after whom a comet is named

60. The English translation for the french word: frangin

61. The best is ___ come'

64. Feature of many modern computer monitors

66. The English translation for the french word: glorifier

67. Eastern European pork fat dish

68. Some police officers: Abbr.

69. Sleuths, for short

70. QB Detmer and others

71. Neckcloth

DOWN

1. Work ___

2. Writer Asimov

3. Wobbly walker

4. Stephen Jay ___, author of "The Panda's Thumb"

5. What some surfers do

6. In ___ (worked up)

7. XX times VIII

8. Fill in the blank with this word: ""___ approved" (motel sign)"

9. The other way around

10. Young lady of Sp.

11. My Heart Can't Take ___ More' (1963 Supremes song)

12. Racine tragedy

14. Palme ___ (Cannes award)

16. Three-time Indy winner Wilbur ___, who introduced the crash helmet

20. Barges

25. The English translation for the french word: dÈpanneur

26. Czar known for his mental instability

27. Fill in the blank with this word: ""Norma" librettist Felice ___"

28. Vitamin whose name sounds like a bingo call

29. French physicist noted for research on magnetism (born in 1904)

30. The English translation for the french word: antipsychotique atypique

31. Year in the middle of this century

32. Filming process for multiple aspect ratios

34. Yard sale tag

35. Keats's "Ode on a Grecian ___"

37. Night ___

42. The "S" of R.S.V.P.

44. Soprano Renata

47. TV announcer Hall

49. The English translation for the french word: tondre

51. It's a relief

52. Some idols

53. Fill in the blank with this word: ""Li'l ___" (Al Capp strip)"

54. Fill in the blank with this word: "___ Hall (Robert Southey's home)"

55. Short answers?

56. Fill in the blank with this word: "___ peace accord (1998 agreement)"

57. Words

59. Fill in the blank with this word: ""The Bells ___ Mary's""

62. Tic-___ (metronome sound)

63. Suffixes with glycer- and phen-

65. Military asst.

PUZZLE 62

ACROSS

1. Race car gauges, for short

6. British verb ending

9. Switch suffix

13. Thrown for ___

14. Verse starter?

15. Physicist Ohm

16. Grant's first secretary of state ___ Washburne

17. The 21st, e.g.: Abbr.

18. Skater Brian

19. It enters things

21. "Drop City" novelist, 2003

23. Old bird

24. Fill in the blank with this word: "___ noche (tonight, in Tijuana)"

25. Overseas bar deg.

28. White Sands Natl. Monument state

30. Bankbook ID

35. See 11-Down

37. The English translation for the french word: Èco

39. Unpretentious instrument

40. WB sitcom

41. Wreckage

43. With 69-Across, 1930s-'50s bandleader

44. Unscramble this word: aradw

46. make amends for

47. Not give ___

48. "The Ecstatic" rapper

50. Showy flower of the iris family

52. Fill in the blank with this word: "Carolina ___"

53. Drift

55. One of TV's "Bosom Buddies"

57. Pal

61. The lonely goatherd, in a "Sound of Music" song

65. "American Beauty" hero

66. Old New Yorker cartoonist Gardner ___

68. They greet each other by pressing their noses together

69. Mystery writer Gardner et al.

70. Having no bounds: Abbr.

71. Cold

72. Fill in the blank with this word: "Erymanthian ___, fourth labor of Hercules"

73. Sportscaster Allen

74. Some Dodges

DOWN

1. Six-foot vis-

2. Pas ___ (gentle ballet step)

3. San Francisco's ___ Tower

4. Yawn-inducing

5. Encourage

6. See 28-Down

7. Show obeisance

8. Half-___ (pipsqueaks)

9. Fill in the blank with this word: "___ Aarnio, innovative furniture designer"

10. Fill in the blank with this word: ""Dawn of the ___ fingers ...": The Odyssey"

11. Turgenev's birthplace

12. You can't help but love Shrek even though he's a big, green one of these monsters

15. Revert

20. Whistle-blower, e.g

22. Windy City transportation org.

24. Evidence in court

25. West Coast N.F.L.'er

26. Pot creators

27. Some cakes

29. S. American land

31. Fill in the blank with this word: "___ Grande, Ariz."

32. French poet (born in Romania) who was one of the cofounders of the dada movement (1896-1963)

33. Roberto Duran's uncle?

34. Zing

36. Unscramble this word: rayd

38. Variety of agate

42. Watch word

45. White Label Scotch maker

49. Zine reader

51. Army medic

54. Round table

56. Norman Vincent ___

57. Not an aristocrat

58. Old magazine ___ Digest

59. Venezuela's ___ Margarita

60. Visionary

61. The American Heritage Dictionary calls this pronoun the most famous feature of Southern dialects

62. The English translation for the french word: logotype

63. The Isle of Man's Port ___

64. Relieves (of)

67. WSW's reverse

PUZZLE 63

ACROSS

1. Parts of masks

6. Hairdo

10. The English translation for the french word: nuque

14. Tacitly agree with

15. Sci-fi princess

16. Writing on the wall

17. -

18. Fill in the blank with this word: "___ brat"

19. Island in French

Polynesia

20. To astronomers, they're hot and blue

22. Jazz's ___ Fleck and the Flecktones

24. Would you mind if we took this inside?'

25. Vetoes

27. Tree whose two-word name, when switched around, identifies its product

29. The English translation for the french word: Ègaliser

33. Round-the-world traveler Nellie

34. Where chamois and snow leopards live: Abbr.

35. Passion

37. Wrap up

41. R&B singer with the hit 'It's All About Me'

42. The Earl of this began shipping some of the marble work from the Parthenon back to England

44. Ham on ___

45. "Ah, Wilderness!" mother

48. Fill in the blank with this word: "___-majestÈ"

49. George Ade's "The Sultan of ___"

50. Fill in the blank with this word: "___ supra (text comment)"

52. Slickers and the like

54. State with the least populous capital

58. Side-channel, in Canada

59. U.S. dance grp.

60. Votes against

62. Soviet agcy. in Bond novels

66. It's the 4-letter term for the thin sheets of dried seaweed in which sushi is wrapped

68. Fill in the blank with this word: "___ point (embroidery stitch)"

70. Wing: Prefix

71. Scottish rejections

72. Wriggling

73. Tests by lifting

74. They go by the wayside

75. Series of legis. meetings

76. Waiting in the wings

DOWN

1. Madrid month

2. Hebrew letters

3. Work in the cutting room

4. Select smokes

5. Unsaturated alcohol

6. Where St. Pete is

7. Subject follower

8. Ditto alternative

9. 1960's-70's antidiscrimination movement

10. Fill in the blank with this word: "___'wester"

11. Unicellular organism

12. Fill in the blank with this word: "___ onion"

13. Vast, old-style

21. Unscramble this word: iesez

23. "Now you're talking!"

26. George ___, longtime maestro of the Cleveland Orchestra

28. Joe-___ weed (herbal remedy)

29. Single-named supermodel

30. Amts.

31. Some, to Spaniards

32. Zealous

36. Old Apple computers

38. Rock's Motley ___

39. The Clan of the Cave Bear' heroine

40. Fear, to Fran

43. Ruhr refusals

46. Elementary suffix

47. The English translation for the french word: ÈbÊne

49. The English translation for the french word: adoucir

51. Since way back when

53. a woman with abnormal sexual desires

54. Woman of letters?

55. W.W. II vessel

56. Prelate's title: Abbr.

57. David who caught a key pass in the 2008 Super Bowl

61. Short answers?

63. Torn (from), old-style

64. Young lady of Sp.

65. The "H" in "M*A*S*H": Abbr.

67. Vol. 1, No. 1, e.g

69. Fill in the blank with this word: "___ admin"

PUZZLE 64

ACROSS

1. V preceder

5. Town line sign abbr.

9. Old draft deferment category for critical civilian work

13. How football's Jerry was addressed as a boy?

16. Lux. neighbor

17. Manages to get through

18. Writer's Market abbr.

19. Time-honored Irish cleric, for short

20. Wiped out

22. Wormer, say

23. Lived ___ (celebrated)

25. Square, in 1950s slang, indicated visually by a two-hand gesture

27. Travelers' papers

30. Stamps, say

32. Sue Grafton's "___ for Alibi"

33. Scattered, as seed

34. Wayne LaPierre's org

35. It may be blacked out

38. Maritime CIA

39. Star born Frederick Austerlitz

41. Lb. or oz.

42. The English translation for the french word: posada

44. Fill in the blank with this word: "Feather ___"

45. Title page?

46. Fill in the blank with this word: ""The ___" (Uris novel)"

47. With 100-Across, Naples opera house Teatro di ___

48. Etc. and ibid., e.g.

49. Declare, old-style

51. He played Mowgli in "Jungle Book"

53. Indonesia's ___ Islands

54. Asian goat

56. Fill in the blank with this word: ""I Am ... ___ Fierce," #1 Beyonc"

59. The Godfather' co-star

61. Apparatus named for a French physician

64. This instrument of the cult of Apollo lent its name to the type of poetry it accompanied

65. Exhibit artfulness

66. "As we have therefore opportunity, let ___ good to all men": Galatians

67. Undesirable serving

68. Way from Syracuse, N.Y., to Harrisburg, Pa.

DOWN

1. White House inits.

2. Writer's supply: Abbr.

3. U.S.A.F. NCO

4. Perfect but impractical

5. Verdi baritone aria

6. Veracruz Mrs.

7. Small songbirds

8. Some batteries

9. USA alternative

10. Tries to trap something

11. TV's ___ twins

12. Time for potty training, maybe

14. The Louvre's Salles des ___

15. Go-aheads

21. Zigzag

24. Parallel ____ Moresby

26. Bookie's charge, for short

27. Letters on a R

28. Prefix with sphere

29. Protection against rustling?

31. Most sacred building in Islam

34. Wiretapping grp.

35. Fill in the blank with this word: "___-la-la"

36. Hate or fear follower

37. Waiting area announcements, briefly

39. Like some professors

40. Water's conductivity comes from these particles, like positive sodium ones & negative chlorine ones

43. Fill in the blank with this word: ""___ approved" (motel sign)"

45. Black & Decker offering

47. Straight run for skiers

48. Lack of restraint

49. Strongly hopes

50. Unilever brand

52. There's many ___ 'twixt the cup and the lip'

53. Since 1920 the organization known by these 4 letters has helped defend the rights & freedoms of our people

55. Cambodian money

57. School subj.

58. Handle: Fr.

60. Fill in the blank with this word: "___-noir (modern film genre)"

62. Band with the 1999 hit "Summer Girls"

63. Width measure

PUZZLE 65

1	2	3	4		5	6	7	8		9	10	11	12	13
14					15					16				
17					18					19				
20				21					22					
23				24					25					
		26				27	28				29	30	31
32	33	34				35					36			
37					38					39				
40					41					42				
43				44					45					
		46					47				48	49	50
51	52	53				54	55				56			
57					58					59				
60					61					62				
63					64					65				

ACROSS

1. Unscramble this word: nrbu

5. Davy Jones or any other Monkee

9. Thurmond who left the Senate at age 100

14. Swift Malay boat

15. Unscramble this word: sfae

16. "Butter knife" of golf

17. Fill in the blank with this word: ""Mi casa ____ casa""

18. Les …tats-____

19. Space ____

20. Acupuncture, e.g.

23. Verdi's "____ giardin del bello"

24. Fill in the blank with this word: "____ Farrow, Mrs. Sinatra #3"

25. Kellogg's Cracklin' ____ Bran

26. Measurements overseer: Abbr.

27. Dusseldorf donkey

29. Fill in the blank with this word: "Dash ____ (write quickly)"

32. Madrid's ____ del Prado

35. Maj. Houlihan portrayer in "M*A*S*H"

36. Golf's ____ Aoki

37. Middle of the quote

40. Thumbs-up responses

41. Former Georgia senator Sam

42. Vacuum cleaner parts

43. Slant

44. Sharp-billed diver

45. Fill in the blank with this word: ""You're the ___" (Cole Porter classic)"

46. Rock's ___ Speedwagon

47. Samuel L. Jackson's character in "Pulp Fiction"

48. Supermarket with a red oval logo

51. 100-Across, for one

57. ___ jazz (fusion genre)

58. Fill in the blank with this word: "___ Morris, signature on the Declaration of Independence"

59. Yours: Fr.

60. Pianist Claudio

61. With a clean slate

62. Barefaced

63. Passed out

64. Fill in the blank with this word: "___ list"

65. Book in which the destruction of Samaria is foreseen

DOWN

1. Rounded end

2. Star bears

3. Martini's partner

4. Suffix with aqua

5. Woes, to a Yiddish speaker

6. Woman of letters?

7. Fill in the blank with this word: ""___ not back in an hour...""

8. Fill in the blank with this word: "___-majestÈ"

9. Unscramble this word: asicol

10. Boston Harbor event precipitator

11. Fill in the blank with this word: ""___, Pagliaccio" (aria)"

12. Opera conductor Daniel ___

13. Speck of dust

21. Form into an arch, old-style

22. Fill in the blank with this word: ""A merry heart ___ good like a medicine": Proverbs"

26. Trawling equipment

27. "Dallas" family name

28. Unscramble this word: sgin

29. She-bears, south of the border

30. #1 thing

31. They're not for you

32. They have open houses

33. Fill in the blank with this word: "___, skip and jump away"

34. The Open Window' author

35. X-rated stuff

36. Waffle House alternative

38. Short concluding stanza

39. Nat King Cole's "___ Things Money Can't Buy"

44. Deleted

45. How "12" is expressed in Chinese

46. Fill in the blank with this word: "___ calculus (kidney stone)"

47. Were in accord

48. Specks

49. Thou

50. The English translation for the french word: akita inu

51. The English translation for the french word: perle

52. Sein : German :: ___ : French

53. Dragsters' org.

54. Supporter's suffix

55. Up ___ good

56. "Happy Days" fellow

PUZZLE 66

1	2	3	4	5	6	7		8	9	10		11	12	13
14								15				16		
17														
18						19				20				
21			22		23				24					
			25	26				27				28	29	30
31	32	33					34					35		
36				37		38				39		40		
41					42				43		44			
45				46				47						
		48	49				50				51	52	53	54
55	56					57						58		
59					60				61	62	63			
64														
65				66				67						

ACROSS

1. Restless, to Rachmaninoff

8. Where I's cross?: Abbr.

11. N.B.A.

14. Was hot

17. Mush room?

18. Suffix with hotel

19. There: Lat.

20. Tallow sources

21. Young lady of Sp.

23. Mao's successor as Chinese Communist leader

24. It gets bigger at night

25. Per ___ (daily)

27. Valueless writing

28. Takes too much, briefly

31. Prepare the way (to)

34. The English translation for the french word: RAU

35. Test in coll., perhaps

36. Portlandia' network

37. "Here's Johnny!" memoirist

40. Drive forward

41. Finish this popular saying: "Waste not want_____."

42. Wall Street org.

43. Not of the cloth

45. Some accounting entries: Abbr.

46. Vice squad?: Abbr.

47. Print tint

48. Keeps

50. Long-running film role

51. Push (down)

55. Gluck's "___ ed Euridice"

57. Literary monogram

58. Fill in the blank with this word: ""___ Mir Bist Du Schoen" (1938 hit)"

59. Stranded golf pro?

64. "What a shame your footwear is missing," palindromically

65. Regular: Abbr.

66. Youngster

67. Thrust

DOWN

1. Upbeat, in music

2. Wowed eyewitness

3. Fill in the blank with this word: ""___ economy is always beauty": Henry

James"

4. Washington State's Sea-___ Airport

5. Wore away

6. Racecar driver ___ Fabi

7. Fill in the blank with this word: "___-gatherum"

8. Obi-Wan, for one

9. The ___-Mags (classic punk rock band)

10. Returns to base

11. "If today is the 18th, then Christmas is next week"?

12. Worry

13. Some wines

15. Sunscreen label abbr.

16. Encouraging sign

22. Word on a dipstick

23. Swingers

24. Shortens a sentence, perhaps

26. Elementary suffix

27. Oom-___ (polka rhythm)

29. Fill in the blank with this word: "Emmy-winning actress ___ de Matteo"

30. Unscramble this word: lsle

31. Fill in the blank with this word: "___ Hayes of "The Mod Squad""

32. Give an ___ effort

33. Floods and such

34. Trucial States, today: Abbr.

38. Misery

39. Power ___

44. Fill in the blank with this word: ""Am ___ risk?""

46. The English translation for the french word: anorak

47. Old-time entertainer ___ Tucker

49. Hall-of-Fame coach Ewbank

50. "@#$%!," e.g.

52. Fill in the blank with this word: "___ to mankind"

53. Red Bordeaux

54. The English translation for the french word: piste de ski

55. Some valuable 1920s-'40s baseball cards

56. Read the ___ act

57. Writer Blyton

60. Uganda's ___ Amin

61. Zine staff

62. Used to be

63. U.K. honour

PUZZLE 67

ACROSS

1. Afghani tongue

7. Not opt.

10. Upscale autos

14. Loosens

15. Fill in the blank with this word: "___ minÈrale"

16. Spa sounds

17. Sci-fi character whose name is an anagram of CAROLINA ISLANDS

20. Role for Ingrid

21. Last: Abbr.

22. Where batters eventually make their way to plates?

23. More than just ask

26. QVC alternative

28. Japanese prime minister Taro ___

31. Patient person's tactic

37. Prefix with -fugal

39. One opening a jail door, say

40. Fill in the blank with

this word: ""___ the brinded cat hath mew'd": "Macbeth""

41. Six-Day War participant: Abbr.

42. Pitts of silent film

43. Makes sure something's done

46. One way to field a ball

48. Query, part 2

50. Fill in the blank with this word: "1099-___ (tax form sent by a bank)"

51. The English translation

for the french word: c.‡.c.

52. Day spa offering

54. Linda Ronstadt's "___ Easy"

58. Israeli airport city

60. Times past

64. Jails

68. Without ___ of hope

69. Sorority letters

70. Tony winner Worth and others

71. Like some coats

72. Fill in the blank with this word: "___ possidetis (as you possess, at law)"

73. Classic work by Montaigne

DOWN

1. Woolly-coated dog

2. Fill in the blank with this word: "___-retentive"

3. Union Sq. and Times Sq.

4. Gridder's on-air greeting, maybe

5. Racecar driver ___ Fabi

6. Western N.C.A.A. powerhouse

7. Former Connecticut governor Jodi

8. Not so genteel

9. Sauve ___ peut

10. Wingding

11. Spring time in Lisbon

12. Larrup

13. Workers need them: Abbr.

18. This German car make got its name from the Latin translation of founder August Horch's name

19. Scorch

24. Airport alternative to JFK or LGA

25. Fill in the blank with this word: "___ tai"

27. Fill in the blank with this word: "___ Digital Shorts (late-night comic bits)"

28. Fill in the blank with this word: "___ part (role-plays)"

29. To look, in Leipzig

30. Right turn ___

32. Jerome Kern tune "___ Forget"

33. Mideast's ___ Strip

34. Japanese beer brand

35. The English translation for the french word: mÈson

36. Unleash pent-up emotions

38. Mrs. Addams, to Gomez

41. Like a warm-up exercise, comparatively speaking

44. Tenor Schipa and others

45. WWII intelligence org

46. Whitman's "A Backward Glance ___ Travel'd Roads"

47. Washington Square News

49. Knots

53. Recent additions to la familia

54. Many a holiday visitor / Bandit

55. TV actress Spelling

56. The English translation for the french word: plaie

57. Vodka in a blue bottle

59. It's you! What a surprise!'

61. Fill in the blank with this word: "___ temperature (was feverish)"

62. Whose woods these ___ think...': Frost

63. Sound of a leak

65. Fill in the blank with this word: "___-ray"

66. When H

67. Some Harvard grads: Abbr.

PUZZLE 68

ACROSS

1. Natal native

5. Imprison

10. The English translation for the french word: piquant

14. Fill in the blank with this word: "Elvis ___ Presley"

15. Tidal points

16. Western Electric founder ___ Barton

17. Source of pop-ups?

20. Ransom ___ Olds

21. Lagerl^f's "The Wonderful Adventures of ___"

22. The English translation for the french word: toner

23. What ___!' ('That price is great!')

24. Aircraft carrier

26. Skin damager, for short

29. Sticky stuff

30. Was taken in

33. This word for a dolt

was the first name of Rube Goldberg's comic strip hero McNutt

34. Where Prince Philip was born

35. Fill in the blank with this word: "___ 1 (Me.-to-Fla. highway)"

36. Aspirant's motto ... or, phonetically, what 18-, 23-, 47- and 57-Across each consist of

40. Word on a dipstick

41. Fill in the blank with this word: "___-fatty acid"

42. The Golden Age of Roman literature runs from Cicero to this "Art of Love" author

43. Scotland's Firth of ___

44. The Clan of the Cave Bear' heroine

45. Title name in a 1965 #1 hit

47. Fill in the blank with this word: ""Mona ___""

48. They go by the wayside

49. Make ___ of (embarrass)

52. Golf's ___ Aoki

53. Fill in the blank with this word: ""There's No Place Like ___" (old TV slogan)"

56. Person eligible for casting?

60. "Now you're talking!"

61. Near Eastern inn

62. Insurance giant

63. Fill in the blank with this word: "___ club"

64. B. & O. stop: Abbr.

65. Caught some rays

DOWN

1. Writer Grey

2. Russia's ___ Mountains

3. Pierre who wrote "P

4. Verse starter?

5. Fill in the blank with this word: ""Annales" poet Quintus ___"

6. The English translation for the french word: grÈmilleux

7. Debutantes' affairs

8. The English translation for the french word: appli

9. Mandela's land: Abbr.

10. Surface again, as a road

11. Have ___ with (know well)

12. Finish this popular saying: "If you build it they will_____."

13. Fill in the blank with this word: "Battle of the ___, opened on 10/16/1914"

18. Hanging ___ a thread

19. Put a cork in

23. Rubaiyat' rhyme scheme

24. Wades through

25. Whether firm, soft or silken, this soy product is an excellent source of protein

26. Wolf pack member

27. Sci-fi author McIntyre

28. The English translation for the french word: chahuteur

29. Never ___ Give You Up' (1988 #1 hit)

30. The Real Housewives' network

31. Fill in the blank with this word: ""Let ___!" ("Go ahead!")"

32. Woman's undergarment

34. Occult science

37. Wakes thrown up behind speedboats

38. W.C.T.U. members

39. Tries to win

45. World Cup match venues

46. Ukrainian city near the Polish border

47. Vouvray wines come from this valley noted for its chateaux

48. Take ___' (host's request)

49. Limp as ___

50. Tumbled

51. To do this is to stare at someone desirously

52. Some flawed mdse.

53. Coordinate in the game battleships

54. Nota ___

55. toward the mouth or oral region

57. Original Dungeons & Dragons co.

58. Most miserable hour that ___ time saw': Lady Capulet

59. Western treaty grp.

PUZZLE 69

1	2	3	4		5	6	7	8	9		10	11	12	13
14					15						16			
17			18								19			
20				21						22				
		23				24		25						
26	27	28			29		30				31	32	33	
34				35		36		37						
38			39		40		41		42					
43			44		45		46		47					
48			49		50		51							
		52			53		54							
55	56	57			58		59				60	61	62	
63				64						65				
66				67						68				
69				70						71				

ACROSS

1. Yorick's skull, for one

5. Violent behavior, to Brits

10. Put a tree after "H" & you get this nautical wheel

14. Now see ___!'

15. Major defense contractor

16. Dilbert co-worker

17. Not delayed

19. Love ___

20. Le coeur a ___ raisons...': Pascal

21. Song that people flip for?

22. Witch of ___

23. Wiretapping grp.

24. Noted Presidential loser

26. Yeah, sorry'

30. Riviera resort

34. Prix de ___ de Triomphe (annual Paris horse race)

35. 1942-45 stats disseminator: Abbr.

37. Pal, in slang

38. Thin as ___

40. Female rap trio with the #1 hit "Waterfalls"

42. U.R.L. opener indicating an additional layer of encryption

43. The English translation for the french word: risquÈ

45. Watership down?

47. Yarn

48. State

50. Work on a galley

52. Shoe brand that sounds like a letter and a number

54. Mother ___

55. Ran over

58. Jewelry chain

60. Fill in the blank with this word: ""O Sole ___""

63. Ormandy's successor in Philadelphia

64. Old medicine?

66. See 103-Across

67. Syndicated astrologer Sydney

68. Western Indian

69. Fill in the blank with this word: ""___ Say," 1939 #1 Artie Shaw hit"

70. Worry greatly

71. Squire

DOWN

1. Letters from Greece

2. Oscar-winning French film director ___ Cl

3. ___ Island (location near Portland, Maine)

4. The Carolinas' ___ Dee River

5. Words starting a simple request

6. Partied, so to speak

7. Peer Gynt Suite' composer

8. Tough guys

9. Grand ___ Opry

10. Scribbled, e.g.

11. Dusseldorf donkey

12. Finish this popular saying: "He who hesitates is___."

13. Business school subj.

18. Made a tax valuation: Abbr.

22. -

23. [See blurb]

25. Unscramble this word: alp

26. Widen

27. Fill in the blank with this word: "___ to go"

28. You ___ right!'

29. Windsor's prov.

31. Fill in the blank with this word: ""CÛmo ___?""

32. Sugar ___

33. Word go

36. Big sizes, briefly

39. Kirsten of "Spider-Man"

41. Studio shout

44. Motel freebie

46. From memory

49. Skin condition

51. Younger brother, say

53. Suspect foul play

55. Struck, once

56. Actor Willard of "The Color Purple"

57. Spillane's '___ Jury'

59. Peseta : Spain :: ___ : Italy

60. Tiny bit

61. Windows picture

62. Kind of romance between actors

64. Stuff

65. Yankee Maris, informally

PUZZLE 70

ACROSS

1. See 38-Down

6. Fill in the blank with this word: ""That suits me to ___": 2 wds."

10. Location

14. "Butter knife" of golf

15. Prefix with sphere

16. Fill in the blank with this word: "___ way, shape or form"

17. Title of some 2004 Summer Olympics preview shows

20. WWW addresses

21. Tent furniture

22. Fill in the blank with this word: "Bel ___"

23. One of the Chaplins

24. Michael of 'Juno'

25. 1969 Beatles hit

26. Soldier

27. Traveler

29. The English translation for the french word: …os

32. 2000 U.S. Open winner

35. Hippie gathering of a sort

36. The Sun and Mercury are in it: Abbr.

37. Kew

40. Mai ___ (drinks)

41. Warm-up for a marathon

42. Was sycophantic to

43. Wilt

44. Make ___ (get paid well)

45. Fill in the blank with this word: "___ Offensive"

46. Suspension

48. Fill in the blank with this word: ""Deutschland ___ Alles""

50. Litmus bluer: Abbr.

53. When to hear "O Romeo, Romeo! wherefore art thou Romeo?"

55. Warm-up for the college-bound

56. Wildcats' org.

57. What boxer #2 was

60. Touched the tarmac

61. Fill in the blank with this word: "___ Lemaris, early love of Superman"

62. Moth-___

63. Wise one

64. Fill in the blank with this word: ""Divine Secrets of the ___ Sisterhood""

65. Silents star ___ Bara

DOWN

1. Them, essentially

2. Second-largest moon of Uranus

3. Manipulate

4. Some airport data: Abbr.

5. Fill in the blank with this word: "___-hoo"

6. King of old movies

7. Making of handicrafts, say

8. Western Electric founder ___ Barton

9. Taxonomy suffix

10. Part of a drum kit

11. Ups and downs of exercise?

12. London's ___ of Court

13. Mateus ___

18. Tennis whiz

19. Formal hat, informally

24. Walking stick

25. Robt. E. Lee, e.g.

26. Relieves (of)

28. The English translation for the french word: groin

30. Poulenc's "Sonata for ___ and Piano"

31. Unscramble this word: dsan

32. Three-stripers: Abbr.

33. Tiny battery

34. Clown's over-the-top topper

35. Popular vacation locale

36. PBS station with a transmitter on the Empire State Building

38. Wrapper abbr

39. Fill in the blank with this word: "___ and terminer"

44. Mideast leader?

45. Mal de ___

47. ___ a stinker?' (Bugs Bunny catchphrase)

49. State in Brazil

50. Well-tuned

51. Like an Interstate

52. "Same goes for me"

53. Fill in the blank with this word: ""An apple ___...""

54. U.S.A.F. Academy site

55. Swift Malay boat

56. Wyle of "ER"

58. West Coast brew, for short

59. This, to Th

Solutions

Puzzle Solution 1

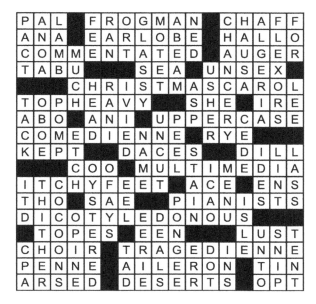

A	C	C	R	A		A	C	H	Y		P	E	R	I
S	H	O	E	D		F	L	O	E		A	R	U	M
H	O	W	D	O	Y	O	U	D	O		P	E	E	P
Y	I	P		B	O	O	B		M	A	Y			
	R	U	P	E	R	T		B	A	R	R	A	G	E
	N	O	S	E		S	E	N	T	I	E	N	T	
P	A	C	K		W	E	A	R	Y		R	A	H	
A	C	H	E		D	A	N	D	Y		M	O	R	E
T	O	E		S	I	S	S	Y		U	S	S	R	
I	R	R	I	T	A	T	E		H	A	L	O		
O	N	S	T	A	G	E		L	A	B	E	L	S	
		A	B	O		S	I	L	L		B	U	R	
P	A	I	L		N	E	T	T	L	E	S	O	M	E
U	G	L	I		A	M	A	H		S	O	M	M	E
T	A	L	C		L	U	G	E		T	U	B	A	L

Puzzle Solution 2

P	A	L		F	R	O	G	M	A	N		C	H	A	F	F
A	N	A		E	A	R	L	O	B	E		H	A	L	L	O
C	O	M	M	E	N	T	A	T	E	D		A	U	G	E	R
T	A	B	U			S	E	A		U	N	S	E	X		
			C	H	R	I	S	T	M	A	S	C	A	R	O	L
T	O	P	H	E	A	V	Y		S	H	E		I	R	E	
A	B	O		A	N	I		U	P	P	E	R	C	A	S	E
C	O	M	E	D	I	E	N	N	E		R	Y	E			
K	E	P	T		D	A	C	E	S			D	I	L	L	
		C	O	O		M	U	L	T	I	M	E	D	I	A	
I	T	C	H	Y	F	E	E	T		A	C	E		E	N	S
T	H	O		S	A	E		P	I	A	N	I	S	T	S	
D	I	C	O	T	Y	L	E	D	O	N	O	U	S			
	T	O	P	E	S		E	E	N			L	U	S	T	
C	H	O	I	R		T	R	A	G	E	D	I	E	N	N	E
P	E	N	N	E		A	I	L	E	R	O	N		T	I	N
A	R	S	E	D		D	E	S	E	R	T	S		O	P	T

Puzzle Solution 3

M	A	N	N	■	S	T	R	A	T	H	■	A	P	S	I	S
A	N	O	A	■	U	R	A	N	I	A	■	P	R	A	N	A
D	O	W	N	H	E	A	R	T	E	D	■	H	E	M	A	L
A	P	A	C	E	■	W	E	E	P	■	R	E	T	I	N	A
M	I	N	E	R	A	L	■	■	I	S	O	L	A	T	E	D
S	A	D	■	■	C	E	M	E	N	T	M	I	X	E	R	S
■	■	T	S	A	R	D	O	M	■	Y	E	A	■	■	■	■
A	D	H	E	R	E	■	O	P	A	L	■	■	B	A	A	L
P	I	E	C	E	D	E	R	E	S	I	S	T	A	N	C	E
T	E	N	T	■	■	P	E	R	I	■	A	E	R	A	T	E
■	■	■	F	I	E	■	O	D	D	N	E	S	S	■	■	■
B	E	A	T	A	R	E	T	R	E	A	T	■	■	T	S	P
A	D	M	O	N	I	S	H	■	■	R	A	I	N	I	E	R
S	E	R	I	N	S	■	R	E	E	K	■	N	E	G	R	O
S	M	I	L	E	■	M	I	D	D	L	E	N	A	M	E	S
E	A	T	E	R	■	B	L	A	D	E	S	■	R	A	N	I
T	S	A	R	S	■	A	L	M	O	S	T	■	S	T	E	T

Puzzle Solution 4

S	O	A	K	■	A	B	R	A	H	A	M	■	R	P	M
H	I	L	A	R	■	L	O	O	S	E	L	Y	■	E	R	A
A	L	P	H	A	■	B	E	S	P	R	I	N	K	L	E	S
G	R	A	N	D	J	U	R	Y	■	■	A	N	I	S	E
S	I	C	■	D	O	M	■	■	H	A	S	H	O	V	E	R
■	G	A	E	L	S	■	A	T	I	L	T	■	W	E	T	S
■	■	■	S	E	S	A	M	E	S	E	E	D	■	■	■
E	R	A	S	■	■	B	E	N	■	R	U	S	S	I	A
V	I	D	E	O	C	O	N	F	E	R	E	N	C	I	N	G
E	M	O	N	E	Y	■	O	V	A	■	■	O	R	S	O
■	■	R	A	T	T	L	E	T	R	A	P	■	■	■
E	G	G	S	■	N	E	E	D	S	■	O	R	E	A	D
P	A	R	T	I	S	A	N	■	■	B	A	R	■	R	I	O
I	L	I	A	C	■	■	P	I	O	N	E	E	R	E	D
C	O	P	Y	H	O	L	D	E	R	S	■	A	V	I	S	O
A	R	P	■	O	N	E	I	R	I	C	■	R	I	V	E	R
L	E	E	■	R	E	D	D	I	S	H	■	■	L	E	S	S

Puzzle Solution 5

Puzzle Solution 6

Puzzle Solution 7

Puzzle Solution 8

Puzzle Solution 9

Puzzle Solution 10

Puzzle Solution 11

Puzzle Solution 12

Puzzle Solution 13

Puzzle Solution 14

Puzzle Solution 15

T	A	P	P	E	T		S	L	A	P	U	P		S	A	E
A	M	R	I	T	A		I	O	N	I	Z	E		M	P	G
L	I	E	C	H	T	E	N	S	T	E	I	N		A	P	R
U	N	M	A	N		S	K	E	W			T	U	L	L	E
S	E	E		I	M	P		R	E	T		I	S	L	E	T
		D	E	C	E	I	T		R	E	A	M	S			
S	L	I	T		N	E	W	S	P	A	P	E	R	M	A	N
P	U	T	U	P		D	O	H		K	E	N		A	L	I
O	N	A	I	R		S	A	T			T	E	N	O	N	
O	A	T		O	D	E		H	I	V		O	R	I	O	N
F	R	E	U	D	I	A	N	S	L	I	P		I	F	F	Y
		R	I	N	S	E		T	R	I	U	N	E			
G	L	O	G	G		E	B	B		I	T	S		S	P	A
N	I	S	E	I		U	R	A	L		E	A	T	E	N	
A	B	C		O	C	C	L	U	D	E	D	F	R	O	N	T
R	Y	A		U	P	R	A	T	E		E	U	G	E	N	E
L	A	N		S	A	T	E	E	N		C	L	O	S	E	D

Puzzle Solution 16

A	D	S		A	N	E	M	I	A	S		B	E	G	E	T
G	E	E		P	A	T	E	L	L	A		A	M	O	U	R
A	L	T	E	R	N	A	T	I	O	N		R	A	N	G	Y
R	I	S	C		H	A	N		A	B	I	D	E			
		H	O	N	E	Y	D	E	W	M	E	L	O	N	S	
B	I	N	O	M	I	A	L		O	E	R		L	I	E	
O	B	I		E	G	G		P	O	K	E	R	F	A	C	E
D	I	S	I	N	H	E	R	I	T		R	Y	E			
E	D	I	T		R	U	S	T	S			A	D	D	S	
		E	W	E		S	T	O	N	E	F	R	U	I	T	
C	O	M	M	I	T	T	E	E		O	R	R		E	R	A
A	R	E		D	U	O		G	O	G	E	T	T	E	R	
M	A	S	H	E	D	P	O	T	A	T	O	E	S			
	T	H	A	N	E		P	A	L			A	I	D	S	
B	O	U	L	E		S	A	L	E	S	P	E	R	S	O	N
U	R	G	E	S		P	L	U	N	K	E	R		L	E	A
D	Y	A	D	S		A	S	S	A	Y	E	R		E	R	G

Puzzle Solution 17

Puzzle Solution 18

Puzzle Solution 19

Puzzle Solution 20

Puzzle Solution 21

Puzzle Solution 22

Puzzle Solution 23

R	A	N	G	■	B	O	D	I	C	E	■	G	R	I	S	T
E	M	I	R	■	A	N	A	L	O	G	■	R	E	N	E	W
D	E	C	I	P	H	E	R	I	N	G	■	A	S	S	A	I
A	B	A	T	E	■	S	E	A	M	■	U	P	H	I	L	L
C	A	R	S	E	A	T	■	■	A	S	P	H	O	D	E	L
T	E	A	■	■	R	E	C	O	N	N	O	I	T	E	R	S
■	■	G	E	N	I	P	A	P	■	A	N	C	■	■	■	■
T	S	U	R	I	S	■	C	H	E	F	■	■	S	L	O	E
S	P	A	G	H	E	T	T	I	J	U	N	C	T	I	O	N
P	A	N	S	■	■	R	I	D	E	■	A	P	A	C	H	E
■	■	■	C	G	I	■	I	C	E	C	U	B	E	■	■	■
M	I	N	E	R	A	L	W	A	T	E	R	■	■	N	B	C
I	S	O	L	A	B	L	E	■	■	L	E	I	S	T	E	R
D	O	N	U	T	S	■	B	A	N	A	L	■	R	A	I	S	E
G	L	A	D	E	■	P	R	O	L	I	F	E	R	A	T	E
E	D	G	E	R	■	R	E	V	O	K	E	■	A	T	O	P
S	E	E	D	S	■	O	R	A	T	E	D	■	N	E	W	S

Puzzle Solution 24

A	C	C	E	P	T	S	■	L	A	M	B	■	I	M	P	S
B	O	O	L	E	A	N	■	A	R	E	A	■	N	U	L	L
U	N	E	M	P	L	O	Y	M	E	N	T	■	S	L	A	Y
T	E	D	■	T	A	R	O	■	O	O	H	■	O	T	T	■
■	■	■	V	I	R	T	U	A	L	R	E	A	L	I	T	Y
M	E	D	I	C	I	■	B	A	A	■	B	E	L	I	E	■
O	Y	E	Z	■	A	L	S	O	■	H	O	E	■	I	N	N
T	E	P	I	D	■	E	U	R	O	■	F	L	I	N	G	S
■	■	R	E	A	T	T	R	I	B	U	T	I	N	G	■	■
S	T	E	R	N	A	■	E	G	I	S	■	A	G	U	E	S
L	A	C	■	D	D	T	■	I	T	E	M	■	L	A	V	E
O	R	I	E	L	■	S	E	N	■	O	M	E	L	E	T	■
T	R	A	V	E	L	A	G	E	N	C	I	E	S	■	■	■
■	A	T	E	■	E	R	E	■	E	A	S	T	■	R	I	G
A	G	I	N	■	M	I	S	S	E	L	T	H	R	U	S	H
L	O	O	T	■	O	N	T	O	■	L	E	O	N	I	N	E
I	N	N	S	■	N	A	S	H	■	A	N	D	A	N	T	E

Puzzle Solution 25

Puzzle Solution 26

Puzzle Solution 27

G	I	L	A	■	C	D	A	G	E	■	E	O	N	S
E	L	I	E	■	R	A	N	O	N	■	L	B	A	R
T	I	N	S	K	I	N	N	E	D	■	N	E	V	A
B	E	D	T	A	B	L	E	■	O	S	I	R	I	S
■	H	N	S	■	T	R	E	N	■					
B	I	E	S	■	E	L	A	S	T	O	M	E	R
T	H	E	T	A	■	N	U	K	E	S	■	A	Y	E
R	A	R	E	■	S	T	E	E	R	■	V	T	E	N
E	N	E	■	I	T	A	G	O	■	Y	O	S	H	I
A	G	I	T	A	T	I	O	N	■	U	N	U	S
■	E	S	E	L	■	I	S	T					
I	M	P	A	I	R	■	W	A	T	E	R	L	O	O
S	O	U	R	■	E	A	R	L	O	F	A	V	O	N
A	L	T	E	■	S	P	I	E	L	■	P	O	R	E
T	E	T	R	■	A	P	T	E	D	■	P	V	T	S

Puzzle Solution 28

O	N	C	E	■	R	U	B	S	■	T	B	S	P
P	A	L	M	■	I	R	A	Q	■	R	U	L	E	R
T	R	O	U	S	S	E	A	U	■	O	N	I	C	E
I	R	A	■	P	E	A	■	A	S	S	I	S	T
M	O	C	H	A	S	■	T	R	I	A	S	S	I	C
A	W	A	Y	■	F	E	E	L	■	F	L	U
■	M	E	D	I	A	■	I	S	S	U	E	R
■	I	N	T	E	R	R	A	C	I	A	L
B	A	L	S	A	M	■	G	N	A	R	L
U	R	L	■	O	K	A	Y	■	S	A	K	I
T	A	F	F	E	T	A	S	■	T	E	A	S	E	R
■	P	A	L	T	E	R	■	C	A	B	■	S	E	A
E	A	T	I	N	■	S	N	O	W	B	O	U	N	D
T	H	E	T	A	■	T	E	R	N	■	F	R	E	E
C	O	D	S	■	S	W	A	Y	■	T	E	D	S

Puzzle Solution 29

Puzzle Solution 30

Puzzle Solution 31

H	A	I	L	S		P	F	U	I			S	N	A	P
U	L	N	A	E		O	O	L	A			I	A	T	E
P	E	R	M	A	N	E	N	T	M	A	G	N	E	T	
S	E	E			G	E	T	Z			U	N	T	E	R
			C	A	V	S		M	E	D	O				
	W	I	L	L	A	C	C	E	L	E	R	A	T	E	
C	O	N	V			H	A	H	N			P	E	S	
A	L	A		S	T	D	E	N	I	S		O	A	T	
I	F	I		A	I	W	A				H	E	R	R	
D	E	R	I	D	E	A	P	R	I	C	O	T	S		
			F	I	R	N		A	D	O	S				
S	G	A	T	E			S	T	A	T		A	P	A	
U	H	O	H	S	P	A	G	H	E	T	T	I	O	S	
F	A	K	E		S	I	T	E			E	E	R	I	E
I	T	I	N		U	S	S	R			N	Y	S	S	A

Puzzle Solution 32

G	I	F	T		E	R	I	C	H			O	O	M
O	L	I	N		S	O	L	D	O		I	M	N	O
L	I	F	T	T	I	C	K	E	T		R	E	D	O
D	A	I		E	D	H	S		W	I	R	I	E	R
			K	N	E	E		L	I	M	E			
C	R	O	O	K	S		F	I	R	E	L	I	N	E
L	O	C	O			I	D	V	E		E	N	E	S
I	T	A	L		A	R	T	I	S		V	O	W	S
N	O	L	A		L	A	D	A			A	U	D	I
T	R	A	I	L	M	I	X		M	O	N	R	O	E
			D	I	A	L		V	O	G	T			
R	A	G	M	A	N		B	A	R	A		C	S	I
C	L	E	A		A	H	A	M	O	M	E	N	T	S
A	T	T	N		C	A	T	O	N		D	E	L	I
F	A	A			S	S	S	S	S		B	T	E	N

Puzzle Solution 33

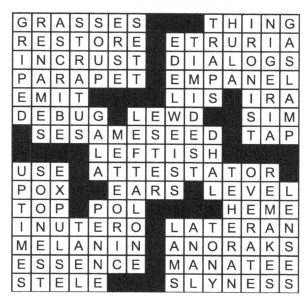

```
C Y T O ■ B O F F O ■ H O K E
N A I F ■ C A R O M ■ E T A T
B Y E S ■ E X U R B ■ A T R A
C A R T I L A G E ■ S R O O T
■ ■ ■ L L C ■ P I S T I L S
M E R G E ■ A G A S S I ■ ■
T I E R R A ■ O W S ■ N C O S
G R U E ■ D B A S E ■ T O R E
S E S E ■ A O L ■ I T H I N K
■ ■ N E T T O N ■ A E R E O
C I R C L E S ■ I B N ■ ■
A S S A I ■ W I N E T H I E F
L A I R ■ S A C E R ■ I M N O
M A D D ■ O N E P M ■ C A L X
A C E S ■ W A R M S ■ S M S H
```

Puzzle Solution 34

```
G R A S S E S ■ T H I N G
R E S T O R E ■ E T R U R I A
I N C R U S T ■ D I A L O G S
P A R A P E T ■ E M P A N E L
E M I T ■ L I S ■ I R A
D E B U G ■ L E W D ■ S I M
■ S E S A M E S E E D ■ T A P
■ L E F T I S H
U S E ■ A T T E S T A T O R
P O X ■ E A R S ■ L E V E L
T O P ■ P O L ■ H E M E
I N U T E R O ■ L A T E R A N
M E L A N I N ■ A N O R A K S
E S S E N C E ■ M A N A T E E
S T E L E ■ S L Y N E S S
```

Puzzle Solution 35

A	S	O	P	■	■	H	A	M	U	P	■	■	S	K	A
O	C	H	O	S	■	E	C	A	S	H	■	■	L	E	M
R	O	S	H	H	A	S	H	A	N	A	■	A	M	I	■
T	R	A	L	A	L	A	S	■	■	R	S	V	P	D	■
A	N	Y	■	M	A	I	■	B	T	A	P	E	■	■	■
■	■	■	M	A	I	D	E	N	V	O	Y	A	G	E	■
A	S	I	A	N	■	■	L	E	A	H	■	N	O	X	■
C	H	M	P	■	T	B	I	R	D	■	A	T	N	O	■
D	I	P	■	B	T	U	S	■	■	R	A	S	E	D	■
C	H	E	F	B	O	Y	A	R	D	E	E	■	■	■	■
■	■	T	T	O	P	S	■	A	A	S	■	E	N	T	■
G	O	U	D	A	■	■	S	I	R	O	C	C	O	S	■
Y	O	O	■	R	A	T	I	N	C	R	E	A	S	E	■
P	R	U	■	D	R	O	M	E	■	B	E	N	E	T	■
S	T	S	■	S	T	O	A	S	■	■	S	S	S	S	■

Puzzle Solution 36

R	O	L	O	S	■	D	R	A	T	■	F	A	R	R
P	R	I	Z	M	■	V	O	U	S	■	A	L	E	A
M	A	S	Q	U	E	R	A	D	E	■	N	L	E	R
S	N	A	G	G	Y	■	N	I	T	■	G	I	V	E
■	■	■	L	E	N	■	T	S	E	T	S	E	S	■
E	N	T	R	Y	F	E	E	■	E	L	A	■	■	■
N	A	B	E	■	U	T	W	O	■	I	S	A	Y	A
N	T	S	B	■	L	W	E	W	O	■	T	R	A	G
A	L	P	E	S	■	T	R	E	S	■	I	A	G	O
■	■	■	L	E	U	■	S	G	P	M	C	M	O	B
W	H	E	R	E	B	Y	■	O	R	O	■	■	■	■
Y	O	R	E	■	U	I	R	■	E	T	A	L	I	A
L	O	E	B	■	R	E	A	D	Y	T	K	I	L	L
I	F	W	E	■	O	L	L	A	■	O	I	L	E	D
E	S	E	L	■	I	D	E	E	■	S	O	T	T	O

Puzzle Solution 37

D	E	P	O	T		Y	V	E	S		A	B	O	O
A	V	O	E	R		O	I	N	K		S	E	P	S
W	A	S	N	O		K	N	E	E		A	S	E	S
		T	O	P	S	E	C	R	E	T	C	O	D	E
L	S	H		H	I	D			I	A	M	S	O	
I	N	A	F	I	X		B	F	L	A	T			
G	O	S	E	E		T	I	T	I		B	S	S	
N	O	T	E	S	O	F	T	H	E	S	C	A	L	E
E	K	E		R	A	E	S		A	H	S	I	N	
		J	E	U	L	R		O	V	I	E	D	O	
A	N	N	A	L		S	U	A		B	E	R		
P	S	I	S	A	N	D	W	E	I	G	H	T		
H	I	T	C		O	R	A	L		E	Y	E	H	S
I	D	A	H		G	A	I	A		L	U	A	U	S
S	E	L	A		S	G	T	S		Y	A	M	A	S

Puzzle Solution 38

M	I	S	C		I	N	C	H	E	D		D	A	D
A	S	K	I		D	E	N	A	D	A		E	N	A
R	A	I	N	C	S	A	N	D	D	S		G	A	R
L	A	B	C	O	A	T		I	S	H	M	A	E	L
E	C	I		W	Y	L	E		R	U	M	I		
E	S	B	A	T		Y	G	O	R		E	L	I	N
			S	O	B		G	H	I	J		L	A	G
	G	P	W	O	D	E	H	O	U	S	E			
S	S	A		N	Y	A	D		S	R	O			
T	I	P	I		D	U	O	S		Y	A	M	A	S
E	G	I	S			N	E	T	B		O	M	I	
A	N	N	A	S	U	I		A	V	O	I	D	E	D
L	O	G		E	L	S	A	M	A	X	W	E	L	L
T	R	L		A	L	P	H	A	D		I	L	I	E
H	A	Y		W	A	Y	A	N	S		N	A	E	S

Puzzle Solution 39

Puzzle Solution 40

Puzzle Solution 41

T	A	R	P		P	S	S	T			S	N	O	B
O	D	O	R		I	C	H	O	R		H	I	R	E
D	A	T	E		N	A	I	R	A		U	S	E	R
D	R	A	C	O		G	R	A	D	A	T	I	O	N
		R	E	N	D		R	H	I	N	O			
S	T	Y	P	T	I	C		S	I	T	U	P	S	
P	E	C	T	O	R	A	L		S	T	R	O	P	P
U	R	L		T	R	I	T	E		O	N	O		
D	R	U	I	D		P	H	Y	S	I	C	A	L	
	A	B	R	O	A	D		Y	A	M	M	E	R	S
	A	T	S	E	A		S	U	P	S				
D	E	A	T	H	T	O	L	L		T	A	S	T	E
A	C	R	E		I	D	I	O	T		S	I	O	N
S	H	U	L		R	A	B	B	I		T	O	L	D
H	O	M	Y		R	I	S	C		O	N	U	S	

Puzzle Solution 42

W	A	S	H		C	H	A	W		G	O	N	Z	O
A	C	H	E		L	A	N	E		A	V	O	I	D
N	A	I	L		O	Y	E	R		R	O	U	N	D
D	I	A	M	O	N	D	W	E	D	D	I	N	G	
		S	W	A	N		R	E	D					
E	N	A	M	E	L		R	U	I	N		A	D	D
N	O	V	A		N	I	S	E	I		D	R	Y	
T	R	A	N	S	L	I	T	E	R	A	T	I	O	N
E	M	S		P	A	C	E	R		R	O	V	E	
R	A	T		I	N	K	S		P	A	U	S	E	S
		A	R	C		D	O	S	S					
	A	N	C	I	E	N	T	H	I	S	T	O	R	Y
G	R	O	U	T		I	R	O	N		F	L	E	E
S	M	O	T	E		B	E	L	T		U	L	N	A
A	S	K	E	D		S	E	E	S		L	A	T	H

Puzzle Solution 43

	T	H	E		F	L	E	W		B	E	I	N	G
P	O	O	R		B	E	M	A		E	N	N	U	I
O	B	N	O	X	I	O	U	S		S	I	T	I	N
P	A	S	S	E					R	I	D	E	S	
U	G	H		B	I	O		F	E	D		R	A	G
P	O	U	L	E	N	C		I	V	E		I	N	N
		A	C	E	T	O	N	E	S		O	C	A	
N	I	G	H		R	O	W	A	N		F	R	E	T
O	N	E		A	R	P	E	G	G	I	O			
A	D	S		T	A	U		L	E	D	G	E	R	S
H	I	T		E	N	S		E	D	O		N	O	T
	G	U	I	L	T					L	A	D	L	E
V	E	R	D	I		M	I	N	U	S	C	U	L	E
A	N	E	L	E		P	R	O	F		T	R	O	D
S	E	D	E	R		H	E	R	O		S	E	N	

Puzzle Solution 44

C	O	Y		C	I	C	A	D	A	S		L	A	M	I	A
U	F	O		G	R	A	N	U	L	E		O	M	E	N	S
S	A	U	D	I	A	R	A	B	I	A		G	E	S	T	S
P	Y	R	E			D	A	G		P	I	N	T	O		
			L	A	R	G	E	I	N	T	E	S	T	I	N	E
R	E	S	I	D	U	U	M		E	S	T		Z	E	E	
A	I	L		E	E	L		M	U	L	T	I	T	A	S	K
G	R	O	U	N	D	P	L	A	N		O	C	A			
S	E	E	K			S	E	P	T	S		I	C	E	D	
		E	V	E		F	L	O	O	R	C	L	O	T	H	
P	A	S	S	I	V	A	T	E		F	U	R		D	N	A
A	S	P		O	A	R			S	I	D	E	R	E	A	L
S	H	O	U	L	D	E	R	B	L	A	D	E	S			
	C	U	R	I	E		H	U	E			V	A	S	E	
B	A	S	I	S		W	I	R	E	T	A	P	P	I	N	G
A	N	E	N	T		I	N	S	T	A	T	E		D	I	G
A	S	S	E	S		T	O	A	S	T	E	R		S	T	Y

Puzzle Solution 45

Puzzle Solution 46

Puzzle Solution 47

S	T	I	R		M	O	D	E	M	S		A	M	A	S	S
A	I	D	E		A	P	O	D	A	L		V	I	L	L	A
P	R	O	L	E	G	O	M	E	N	A		E	D	G	A	R
S	O	L	I	D	E	R		M	I	N	I		N	I	N	E
			C	O	N	T	R	A	C	T	B	R	I	D	G	E
R	A	P			T	O	E			S	I	N	G			
E	R	R	A	T	A		T	E	G		S	A	H	A	R	A
C	R	E	P	U	S	C	U	L	A	R		T	R	O	T	
C	O	M	P	T		G	R	I	M	E		A	S	C	O	T
E	Y	E	R		I	N	D	I	V	I	D	U	A	T	E	
S	O	D	O	F	F		S	E	N		T	O	N	N	E	S
			B	I	L	K			G	S	A			A	R	T
C	L	E	A	R	A	N	C	E	S	A	L	E	S			
H	I	S	T		T	E	A	L		V	I	N	E	G	A	R
O	S	T	I	A		E	P	I	D	I	A	S	C	O	P	E
O	L	E	O	S		L	O	T	I	O	N		T	R	E	F
K	E	R	N	S		S	N	E	E	R	S		S	E	X	T

Puzzle Solution 48

P	R	O	P	H	A	S	E		A	C	I	D		N	O	S
R	E	P	E	A	T	E	R		L	O	C	I		O	V	A
O	C	E	A	N	O	G	R	A	P	H	E	R		N	E	T
O	U	R		P	O	S	H		A	C	T		P	R	E	
F	R	A	N	C		A	M	B	U	S	C	A	D	E		
	A	R	T	S			B	I	B		E	R	I	N		
C	A	R	B	O	H	Y	D	R	A	T	E		L	A	D	S
I	D	O		W	A	N	N	A			T	E	L			
D	O	M	I	N	I	C	A	N	R	E	P	U	B	L	I	C
	A	S	S			C	O	Z	E	N		E	R	A		
E	T	N	A		F	A	T	H	E	R	S	I	N	L	A	W
M	E	T	A		O	B	I		A	O	N	E				
P	R	I	C	E	L	E	S	S			G	O	N	G	S	
O	R	C		M	I	T		O	P	E	D		I	L	O	
W	E	I		B	A	T	T	L	E	C	R	U	I	S	E	R
E	N	S		A	G	A	R		S	H	I	P	M	E	N	T
R	E	M		R	E	L	Y		T	O	P	S	P	I	N	S

Puzzle Solution 49

Puzzle Solution 50

Puzzle Solution 51

A	M	A	H			P	S	I			R	I	S	C
D	I	V	A		P	I	P	S		G	E	N	O	A
A	N	O	N		A	N	A	L		A	L	K	Y	D
R	I	N	G	F	E	N	C	E		G	O	Y	A	
		A	L	L	A	Y		D	E	C				
A	S	T	R	A	L		D	I	S	A	V	O	W	
M	P	H		R	A	T	T	A	N		T	A	R	E
A	R	O	S	E		O	A	T		G	E	L	I	D
Z	I	N	C		G	A	T	E	A	U		V	E	G
E	G	G	H	E	A	D		A	L	L	E	L	E	
		M	A	D		M	A	C	L	E				
	S	C	A	R		P	A	C	H	Y	D	E	R	M
C	H	I	L	L		A	N	T	E		G	A	I	A
F	A	T	T	Y		L	I	O	N		E	S	P	Y
C	H	E	Z			E	A	R			R	Y	E	S

Puzzle Solution 52

L	A	O	A	T		M	O	T	H		C	O	W	L
A	P	N	E	A		A	F	C	A		A	L	L	A
D	E	C	O	N	T	A	M	I	N	A	T	I	O	N
	R	E	L		O	R	E		D	R	H	O	O	K
		U	P	N			U	S	I	A				
	N	U	S	A	N	D	A	N	A	L	Y	S	I	S
P	U	N		R	E	F	L	E	W		A	N	N	O
O	N	I	C	E		L	I	A		O	N	E	G	A
W	C	T	U		L	A	S	S	E	N		V	E	R
S	A	Y	P	R	E	T	T	Y	P	L	E	A	S	
			B	E	T	S		O	Y	L				
S	A	M	O	A	S		B	U	D		I	O	R	
S	U	G	A	R	R	A	Y	L	E	O	N	A	R	D
B	R	E	R		I	O	N	A		O	O	H	E	D
Y	A	R	D		P	N	O	N		P	R	U	D	E

Puzzle Solution 53

Puzzle Solution 54

Puzzle Solution 55

A	R	M		M	A	S	T	I	C	S		S	A	R	A	H
R	U	E		O	P	H	I	D	I	A		E	R	O	D	E
A	B	S	E	N	T	E	E	I	S	M		R	O	O	D	S
B	E	A	U			D	O	C		S	O	M	M	E		
		R	O	C	K	Y	M	O	U	N	T	A	I	N	S	
C	A	R	O	T	E	N	E		S	A	Y		E	D	O	
A	C	E		T	I	E		P	U	E	R	P	E	R	A	L
B	A	N	D	O	L	E	E	R	S		L	E	T			
S	I	T	E		L	A	Y	E	R		C	O	R	A		
		L	A	V		S	E	D	A	N	C	H	A	I	R	
A	R	T	I	F	I	C	E	R		D	I	E		R	P	M
K	O	I		L	O	W		W	A	L	R	U	S	E	S	
A	S	S	A	U	L	T	C	O	U	R	S	E	S			
	S	A	N	T	A		U	P	S			E	A	S	T	
P	I	N	O	T		D	R	E	S	S	C	I	R	C	L	E
A	N	E	L	E		D	I	N	E	O	U	T		M	A	D
R	I	S	E	R		T	O	S	S	U	P	S		E	M	S

Puzzle Solution 56

R	U	T	S		S	H	U	T	S		M	A	K	O
A	B	U	T		P	E	R	I	L		I	C	E	D
M	O	T	E		A	M	E	B	A		S	T	Y	E
P	A	T	R	O	N	S	A	I	N	T	S			
S	T	I	N	G			A	G	A	I	N	S	T	
		A	L	K	I	E		L	L	A	N	O		
O	T	T		E	N	T	R	A	N	C	E	W	A	Y
C	A	R	S		E	E	R	I	E		S	A	F	E
T	W	E	L	V	E	M	O	N	T	H		B	U	D
A	S	S	A	I		R	U	S	E	S				
D	E	S	P	O	T	S		R	I	D	E	S		
		B	L	A	T	H	E	R	S	K	I	T	E	
C	O	L	A		F	R	E	Y	A		K	O	H	L
O	W	E	N		F	I	R	E	D		I	D	O	L
P	L	U	G		Y	A	R	D	S		M	E	S	S

Puzzle Solution 57

Puzzle Solution 58

Puzzle Solution 59

S	L	A	V		F	E	A	T		I	M	B	U	E
N	O	N	E		U	N	D	O		N	A	I	R	A
O	U	T	L	A	N	D	E	R		L	I	N	E	R
T	R	I	A	L	R	U	N		P	A	N	D	A	S
			C	U	E		B	A	W	L				
S	A	L	M	O	N		C	I	G		Y	E	A	H
T	R	O	U	T		O	O	Z	E		L	A	Y	
R	O	O	M	T	E	M	P	E	R	A	T	U	R	E
U	M	P			R	E	S	T		L	E	D	O	N
M	A	Y	A		A	G	E		G	A	L	E	N	A
		M	E	S	A		B	A	R					
U	P	S	I	D	E		M	A	R	M	I	T	E	S
S	O	N	D	E		D	U	S	T	S	T	O	R	M
P	R	I	S	M		A	L	E	E		C	L	I	O
S	E	P	T	A		H	E	R	R		H	U	N	G

Puzzle Solution 60

A	R	C	A		P	A	A	V	O		M	A	N	
N	A	C	R	E		E	M	B	A	R		O	N	E
I	N	C	I	D	E	N	T	A	L	S		N	N	W
M	A	L	A	W	I		A	L	I	A	S	E	S	
		S	H	T	U	P		I	N	S	T	E	P	
T	O	O		I	R	R	E	G		I	B	E		
O	R	R		T	E	S	S	A	S		I	R	K	S
G	U	I	T	E	A	U		V	A	U	G	H	A	N
A	B	E	E		N	L	E	A	S	T		I	O	O
		N	N	E		A	N	G	S	T		T	N	G
M	E	T	A	L	S		S	E	E	E	M			
A	D	A	M	I	T	E		R	R	A	T	E	D	
R	S	T		T	R	A	D	E	S	E	C	R	E	T
L	O	E		E	A	S	E	L		D	E	E	R	E
O	N	S		S	T	T	N	G		S	O	O	N	

Puzzle Solution 61

Puzzle Solution 62

Puzzle Solution 63

E	Y	E	H	S		F	V	M	G		N	A	P	E
N	O	D	A	T		L	E	I	A		O	M	E	N
E	D	I	V	E		A	R	M	Y		R	E	A	O
O	S	T	A	R	S		B	E	L	A		B	R	R
	N	O	E	S		O	I	L	P	A	L	M		
E	Q	U	A	L	I	Z	E		B	L	Y			
M	T	N	S		Z	E	A	L		R	E	C	A	P
M	Y	A		E	L	G	I	N		R	Y	E		
E	S	S	I	E		L	E	S	E		S	U	L	U
	U	B	I		R	A	I	N	W	E	A	R		
V	E	R	M	O	N	T		S	N	Y	E			
A	B	T		N	A	Y	S		S	M	E	R	S	H
N	O	R	I		G	R	O	S		P	T	E	R	O
N	A	E	S		E	E	L	Y		H	E	F	T	S
A	T	V	S		S	E	S	S		O	N	T	A	P

Puzzle Solution 64

R	S	T	U		E	S	T	D		T	W	O	A
M	A	S	T	E	R	R	I	C	E		B	E	L	G
N	E	G	O	T	I	A	T	E	S		S	A	S	E
	S	T	P	A	T		S	L	E	W		V	E	T
		I	T	U	P		L	S	E	V	E	N	
V	I	S	A	S		O	K	S		A	I	S	
S	O	W	N		N	R	A		T	V	G	A	M	E
O	N	I		A	S	T	A	I	R	E		W	G	T
P	O	S	A	D	A		B	O	A		D	E	E	D
	H	A	J		S	A	N		A	B	B	R	S
	P	G	A	C	R	C		S	A	B	U		
A	R	U		T	A	H	R		S	A	S	H	A
C	A	A	N		G	U	I	L	L	O	T	I	N	E
L	Y	R	E		U	S	E	F	I	N	E	S	S	E
U	S	D	O		S	L	O	P		R	T	E	E

Puzzle Solution 65

Puzzle Solution 66

Puzzle Solution 67

```
P A S H T U ■ R E Q ■ B M W S
U N T I E S ■ E A U ■ A A H S
L A N D O C A L R I S S I A N
I L S A ■ U L T ■ I H O P S
■ D E M D ■ H S N ■
A S O ■ W A I T I N G G A M E
C E N T R I ■ R E L E A S E R
T H R I ■ S Y R ■ Z A S U
S E E S T O I T ■ O N A H O P
A N D H I S M O N E Y ■ I N T
T S P ■ O R U B
I T S S O ■ L O D ■ E R A S
L O C K S B E H I N D B A R S
A R A Y ■ L S H ■ I R E N E S
W I R Y ■ U T I ■ E S S A I S
```

Puzzle Solution 68

```
Z U L U ■ E M B A R ■ R A C Y
A R O N ■ N E A P S ■ E N O S
N A T I O N A L P A S T I M E
E L I ■ N I L S ■ T O N E R
A B U Y ■ F T O P
U V R A Y S ■ G O O P ■ B I T
B O O B ■ C O R F U ■ R T E
O N W A R D A N D U P W A R D
A D D ■ T R A N S ■ O V I D
T A Y ■ A Y L A ■ S L O O P Y
L I S A ■ A T V S
A F O O L ■ I S A O ■ H B O
R E G I S T E R E D V O T E R
A L L R ■ S E R A I ■ A E N A
G L E E ■ R R S T A ■ S N E D
```

Puzzle Solution 69

P	R	O	P	■	A	G	G	R	O	■	H	E	L	M
H	E	R	E	■	L	O	R	A	L	■	A	S	O	K
I	N	R	E	A	L	T	I	M	E	■	N	E	S	T
S	E	S	■	S	I	D	E	B	■	S	D	L	T	G
■	■	N	S	A	■	G	O	L	D	W	■	■	■	■
F	R	A	I	D	S	O	■	S	A	N	R	E	M	O
L	A	R	C	■	K	N	X	■	P	A	I	S	A	N
A	R	E	E	D	■	T	L	C	■	H	T	T	P	S
R	I	S	Q	U	E	■	S	U	B	■	T	A	L	E
E	N	O	U	N	C	E	■	T	Y	P	E	S	E	T
■	■	A	S	I	C	S	■	H	E	N	■	■	■	■
S	P	I	L	T	■	Z	A	L	E	S	■	M	I	O
M	U	T	I	■	G	E	R	I	A	T	R	I	C	S
I	G	H	T	■	O	M	A	R	R	■	O	T	O	E
T	H	E	Y	■	E	A	T	A	T	■	G	E	N	T

Puzzle Solution 70

N	G	W	A	Y	■	V	C	E	Z	■	H	K	I	R
O	E	I	R	O	■	I	O	N	O	■	I	N	N	O
T	H	E	R	O	A	D	T	O	A	T	H	E	N	S
U	R	L	S	■	C	O	T	S	■	P	A	E	S	E
S	Y	D	■	C	E	R	A	■	G	E	T	B	■	■
■	■	■	R	A	R	■	G	O	E	R	■	E	O	S
S	A	F	I	N	■	B	E	I	N	■	W	N	B	A
G	A	R	D	E	N	S	I	N	L	O	N	D	O	N
T	A	I	S	■	T	E	N	K	■	Y	E	S	E	D
S	A	G	■	A	W	A	D	■	T	E	T	■	■	■
■	■	H	A	L	T	■	U	B	E	R	■	A	L	K
A	C	T	I	I	■	P	S	A	T	■	N	C	A	A
D	O	W	N	F	O	R	T	H	E	C	O	U	N	T
A	L	I	T	■	L	O	R	I	■	E	A	T	E	N
Y	O	G	I	■	Y	A	Y	A	■	T	H	E	D	A

Made in the USA
Monee, IL
07 November 2020

46961877R00105